Enterprise Information Management in Practice

Managing Data and Leveraging Profits in Today's Complex Business Environment

Saumya Chaki

Apress®

Enterprise Information Management in Practice: Managing Data and Leveraging Profits in Today's Complex Business Environment

ISBN-13 (pbk): 978-1-4842-1219-6

ISBN-13 (electronic): 978-1-4842-1218-9

Managing Director: Welmoed Spahr
Acquisitions Editor: Celestin Suresh John
Developmental Editor: Matthew Moodie
Technical Reviewer: Arnab Naskar
Editorial Board: Steve Anglin, Pramilla Balan, Louise Corrigan, James DeWolf, Jonathan Gennick, Robert Hutchinson, Celestin Suresh John, Michelle Lowman, James Markham, Susan McDermott, Matthew Moodie, Jeffrey Pepper, Douglas Pundick, Ben Renow-Clarke, Gwenan Spearing
Coordinating Editor: Rita Fernando
Copy Editor: Judy Levine
Compositor: SPi Global
Indexer: SPi Global

Distributed to the book trade worldwide by Springer Science+Business Media New York, 233 Spring Street, 6th Floor, New York, NY 10013. Phone 1-800-SPRINGER, fax (201) 348-4505, e-mail orders-ny@springer-sbm.com, or visit www.springer.com. Apress Media, LLC is a California LLC and the sole member (owner) is Springer Science + Business Media Finance Inc (SSBM Finance Inc). SSBM Finance Inc is a Delaware corporation.

For information on translations, please e-mail rights@apress.com, or visit www.apress.com.

Apress and friends of ED books may be purchased in bulk for academic, corporate, or promotional use. eBook versions and licenses are also available for most titles. For more information, reference our Special Bulk Sales–eBook Licensing web page at www.apress.com/bulk-sales.

Any source code or other supplementary materials referenced by the author in this text is available to readers at www.apress.com. For detailed information about how to locate your book's source code, go to www.apress.com/source-code/.

To my guiding lights Bapu, Baba, and Ma who inspired me to write

Contents at a Glance

Contents

About the Author

Saumya Chaki was born in Kolkata, India. He attended the Assembly of God Church School and graduated from Presidency College, both in Kolkata. He obtained his Master in Technology from the Indian School of Mines, Dhanbad. Saumya is an experienced information management consultant with Fortune 500 and Fortune 100 clients from all over the world. As a general manager in the strategy and analytics consulting practice at IBM, he advises CIOs, CMOs, CXOs, and senior architects on information strategy. He has published white papers on information management (formerly DM Review) and SQL Server performance. He is a frequent speaker at numerous information management seminars. His interests include sustainable development, internet of things, and mineral economics. His hobbies include travel, reading, and blogging. He is also the author of the book *A Journey Through 100 Years of Indian Cinema.*

About the Technical Reviewer

Arnab Naskar is resident architect and manager with Accenture's Technology Consulting practice having over 14 years of experience in systems integration consulting. He is a seasoned professional with experience in technical sales, architecture, program management, and service transitioning helping large business conglomerates worldwide in ERP, data management, enterprise and cloud BI, and enterprise big data management and analytics based technology transformations.

Acknowledgments

Writing a book with a busy work schedule, which involved international travel is extremely challenging. This book was conceived over two years ago and I kept delaying, until an opportunity with Apress came in early 2015. Keeping up with the Apress deadlines meant sleeping less and working on weekends and this would not have been possible without the support and encouragement of my wife, Madhu, and my daughter, Shreya. I also would like to thank Rita Fernando, my coordinating editor at Apress, for being so proactive in following up on my progress and reminding me of my deadlines. Rita was my go to person during these seven months and I felt very much at home. I also extend my heartfelt thanks to the entire editorial team at Apress for supporting me all the way through.

I also take this opportunity to thank my acquisition editor, Suresh Celestin John, for having faith in me and guiding me through the book proposal process. Last but not the least, my sincere heartfelt thanks to Arnab Naskar, my ex-colleague and friend for taking time out from his hectic schedule to review and provide his valuable feedback. Arnab is an accomplished architect and it has been a pleasure collaborating with him on numerous assignments, including this book.

Introduction

Every book starts with an idea. The idea behind this book *Enterprise Information Management in Practice* was to write a book, largely from a practitioner's standpoint. An analysis of some of the existing literature concerning enterprise information management showed mainly two trends, scholarly books and technical books. The scholarly books were written by academicians and were consumed in different information management courses in business schools or other master's degree programs. The second trend, technical, written by architects, was of course directed to architects and technical project managers. A third category, a practitioner's perspective is what resulted in this book. The idea was to write in a way in which students as well as professionals could understand, and a conscious effort has been made to avoid using jargons. Concepts have been explained in simple English to make it an easy read. This book would be relevant to a wide audience including—C-level executives, solution architects, project managers, and students of EIM.

Enterprise Information Management in Practice is divided into 13 chapters.

- Chapter 1 explains the definition, scope, and brief history of enterprise information management (EIM).

- Chapter 2 discusses the lifecycle of EIM from creation to destruction.

- Chapter 3 discusses the components of EIM. It is advisable that readers read Chapters 1, 2, and 3 in sequence, after which the remaining Chapters 4 through 11 can be read in any sequence.

- Chapter 4 deals with information sourcing.

- Chapter 5 discusses information integration and exchange.

- Chapter 6 is about information governance and quality and the crucial role it plays in the EIM lifecycle.

- Chapter 7 examines master information management.

- Chapter 8 deals with information warehousing.

- Chapter 9 examines information delivery and consumption.

- Chapter 10 discusses metadata management.

- Chapter 11 examines the emergence of big data solutions and its coexistence with information warehouses.

- Chapter 12 deals with building an EIM solution and the crucial role of an EIM CoE in ensuring optimal solutions.

- Chapter 13 examines some recent EIM trends in today's businesses including big data use cases in mining and metals, oil and gas, and retail.

- Chapters 12 and 13 should be read in the end, once the reader has an understanding of the key solution components.

An effort has been made to keep the chapters short and concise, as the objective was to provide the reader a broad understanding of the topics covered. The content has been made keeping in mind latest developments in EIM space, to provide readers with the latest trends and thinking in the industry including internet of things, big data, and the emergence of cloud-based deployments. The content has both business as well as technical content as the science of enterprise information management involves the understanding of both business and technology. The liberal use of the strategy of execution frameworks is to enable C-level executives to visualize how these concepts can be leveraged in their respective organizations. A few real-world examples also are provided to give a flavor of how EIM solutions are being conceptualized and the business benefits derived. The book would be useful to students of EIM as well as young professionals. Students would obtain insights about how EIM projects are executed and what an EIM reference architecture looks like. Young professionals would have a ready reckoner for the key solution components of EIM. Happy reading and happy learning.

—Saumya Chaki
November 2015

■ ■ ■

CHAPTER 1

Enterprise Information Management: Definition, Scope, and History

Enterprise information management (EIM) is a field of interest specific to the business intelligence and enterprise data warehousing area. It is a field that specializes in finding the optimum use of information assets of an enterprise (both structured and unstructured) to support the decision-making processes as well as managing the performance of an enterprise.

In this chapter, I cover the definition and scope of EIM, including those elements that go into defining an EIM strategy, such as business intelligence strategy, information governance strategy, and others. Also included is a short history of EIM to provide an understanding of how the discipline has evolved with the changing times.

■ **Note**   Enterprise information management is an evolving discipline due to the change of businesses in a digital ecosystem. This calls for new processes and technologies to process the large and diverse volumes of data generated by business in the digital age.

Definition of EIM

A formal definition of *EIM* is the **set of business processes, disciplines, and practices used to manage the information created through an organization's execution of business processes managed by applications and treating this information as an enterprise asset.** Information is truly an enterprise asset that helps organizations execute their business strategy and analyze performance through a pair of leading and lagging indicators. Gartner defined *Enterprise information management* as "an integrative discipline for structuring, describing and governing information assets across organizational and technological boundaries to improve efficiency, promote transparency and enable business insight."

Enterprises today deal with complex business environments in which information demands are given in real time, are complex, and are often the only means to differentiate between competitors. Given this background and the global nature of enterprises the need for effective management of information is crucial in managing enterprises. EIM has evolved as a specialized discipline in the business intelligence (BI) and enterprise data warehousing (EDW) field to address the complex needs of information processing and delivery. EIM deals with both structured and unstructured data. Global enterprises deal with both structured data (e.g. sales data, customer data) as well as unstructured data (e.g. customer satisfaction forms, e-mails, documents, social network sentiments). With the deluge of information that enterprises face given their global operations and complex business models, it is not surprising that making sense of the large amount of data is of paramount importance. More and more, enterprises are investing in the management of information assets to make sense of the data and derive actionable intelligence from the data compiled from business operations.

EIM's key driver is to support the business strategy of the enterprise and to support its business objectives such as profit, revenue, cost optimization, and so forth. At the core of this support is an EIM strategy that details how that data should be integrated, governed, and managed across the information lifecycle of the enterprise. One of the key barriers to EIM is the lack of consistent data definitions and the lack of consistent business rules and jargons that are used across different functional and business units.

EIM's Scope

It is important to understand what goes into EIM and the governing principles that make EIM a specialized branch of data management. Although there are numerous definitions that define the perspectives of EIM, it is important to understand the scope boundaries of EIM as a subject of study and practice. EIM can be defined as management of enterprise information assets both structured and unstructured that provide actionable insights into the operations and performance management of an Enterprise. It also includes the information exchange that happens in a world of collaborative commerce in which enterprise often exchanges crucial BI with partners in the supply chain as well as trading partners and suppliers. There are business models and information exchange standards that define the nature of information exchanged between parties in a collaborative model. It is important to tap into these information exchanges as they provide crucial insights into the operational effectiveness of the business relationships and the success of shared business goals. Hence the scope of EIM is not only limited to the business processes within an enterprise but also business processes related to collaboration and supply chains. This implies that enterprise information can be both internal and external. In addition to such information sources comes the frequent need for market intelligence information (bought from external agencies such as IMS, AC Nielsen, etc.). Hence the need to process many different types of information and their semantic context becomes relevant. Another aspect of EIM is to manage the lifecycle of information from creation to archival. Managing the lifecycle of information implies understanding the parties that create and consume the information, as well as the security needs that surround the information exchange and consumption. Chapter 2 delves in detail the processes of the lifecycle and the associated information management practices.

Having provided a brief background of EIM in the modern business environs, I now move on to a more formal definition of the *scope boundaries* of EIM. To define the scope of EIM it is important to first study the objectives of EIM. The key objective of EIM is to define an information management strategy that looks at the information needs of an enterprise and enables decision making based on integrated enterprise information (both structured and unstructured), which helps in the execution of business vision and strategy. The four key dimensions of EIM for an enterprise are people, process, technology, and infrastructure. To implement an EIM strategy an enterprise needs people in key roles; data management processes that enforce key principles; technologies that support extraction, transformation, cleansing, and storage of data; and a supporting infrastructure that would support the technologies involved in implementing the EIM. The pillars of an EIM strategy, which also define its scope boundaries, would include the following:

- *Business intelligence strategy*

- *Data integration strategy*

- *Master data management strategy*

- *Information governance strategy*

- *Information quality strategy*

- *Data architecture strategy*
- *Enterprise content management strategy*
- *Information security strategy*

I now discuss each of the pillars of an EIM strategy one by one.

Business Intelligence Strategy

BI deals with the information dissemination of enterprise data to business users and senior management that empowers them to make strategic and operational decisions. A BI strategy deals with understanding the enterprise's business objectives and how the existing information landscape caters to the existing business information needs. The "to be" state of an information landscape is often derived from the business objectives moving forward and with help in the assessment of the gaps in the information delivery and dissemination landscape. The BI initiatives are assessed based on the business value they provide to the enterprise and business priorities given to these initiatives. The end result of a BI strategy is a set of BI initiatives based on the current gaps in the information landscape and the roadmap that describes the initiatives which would take the enterprise from the current state to the "to be" state.

Data Integration Strategy

Data integration deals with the integration of enterprise data across applications to build a single consolidated view of enterprise business performance. Data integration strategy deals with the optimal way in which enterprises can build the single consolidated view of business operations and performance. Data integration deals with structured data. More and more enterprises need to integrate unstructured data such as documents, e-mails, and chat logs, which are usually integrated through enterprise content management (ECM) tools. Data integration strategy looks at the best possible integration architecture and the use of re-useable integration components while integrating new sources of data. The integration architecture looks at integration options with source system applications and the most optimal way to pull/push the information into the EDW and downstream data marts. Change-data-capture mechanisms are often used to ensure that only the incremental changes are picked up from the source applications. The objective of a data integration strategy is to ensure that the data integration architecture is optimal in terms of performance and scalability and can meet the enterprise batch window defined for processing data before information delivery processes can kick off (i.e., report schedules can be executed for report delivery to business users).

Master Data Management Strategy

Master data, as the name implies, deals with the key business entities such as product, customer, service, employee, and supplier. Master data management (MDM) strategy helps in delineating the source systems of master data creation and systems that would update/delete and consume master data. The MDM strategy deals with defining an MDM architecture that could be operational/analytical or hybrid depending on the business objectives the strategy would need to address. The MDM strategy also deals with building a business case for MDM that would highlight the tangible business benefits derived. This is an often overlooked opportunity for building a stronger case for MDM implementation. With the rise of global supply chains and more collaboration between retailers and manufacturers, the need for accurate master data is of paramount importance. For enterprises with global supply chains and a need for global data synchronization (GDSN), MDM is a must to have. MDM strategy also defines the MDM data hub and data synchronization needs for the consuming applications based on business needs.

Information Governance Strategy

Another key strategy that deals with the administration of information usage and governance is the information governance strategy (formerly known as data governance). Information governance deals with the crucial aspect of enterprise information usage, consumption, and governance processes and policies concerning this information. With an increasing realization of how information assets are key to the success of an enterprise comes the need for information governance. Information governance strategy deals with classifying enterprise information assets based on usage patterns and business criticality and defining governance structure and policies around the usage and consumption of the information assets. Information governance also brings into focus the quality of the information as the usage and decision making is often impacted with the quality of the rendered information. Hence two key offshoots of information governance strategy are information quality and information security. Depending on the scope of the information assets these can be spun off into separate strategy engagements with the enterprise guidelines of information usage and quality coming from the information governance strategy. Information governance plays a key part in ensuring crucial enterprise data, such as information about customers, product designs, are protected from both internal and external threats. Information governance strategy also helps in realizing the maximum value of enterprise data by opportunities such as effective customer data management to leverage opportunities to up sell and cross sell.

Information Quality Strategy

Information quality strategy can be seen as an offshoot to the information governance strategy. In some cases, information quality strategy can be seen as an independent strategy that ensures enterprise data assets are of optimal quality and strategies are in place to monitor information quality from time to time as well as provide remedies as needed. Often information quality initiatives emerge as part of MDM assessments. Enterprises in need of clean master and reference data need an information quality strategy and processes in place to ensure optimal standards of master and references data entities and attributes. Information quality strategy also must address requirements concerning the quality of data exchanged between enterprises and business partners.

Data Architecture Strategy

Data architecture strategy is one of the key pillars of EIM landscape. Data architecture defines the way data entities are modelled for system of record, data marts, and operational data stores. Data architecture includes the policies and rules concerning how data are sourced, stored, integrated, arranged, and used in decision support systems such as data warehouses, data marts, and operational data stores. Data architecture blueprints also ensure that information management programs align with the key business objectives.

Enterprise Content Management Strategy

Enterprise content management is the strategies, methods, and tools used to capture, manage, store, and deliver documents and content related to organizational processes. For instance, in the insurance industry the policy administration process captures and retains multiple documents about a customer; or in a contracts management process the numerous documents and content, stock performance could be stored. The enterprise content management strategy also controls access to the content to the right people at the right time to empower decision making.

Information Security Strategy

Information security strategy is crucial to protect the enterprise data assets that contain critical information about enterprise business strategies, business performance, and intellectual property. With an increasing number of data breaches occurring, it is of paramount importance for enterprise to have an information security strategy that caters to different types of enterprise data assets. There needs to be a classification of data assets based on sensitivity of the information and information security policies to assess and mitigate the data breach possibilities. Some of the key types of information security threats include identity theft; loss of digital media, such as computer tapes or hard drives; hacking; and so forth.

A Brief History of Enterprise Information Management

EIM started in the early 1990s with the rise of structured data management through data warehouse and data mart implementations. The initial focus was concerned with building decision support systems and the focus of EIM was getting the data in a structured form to the data warehouse or system of record. With the development of relational database management systems such as Oracle, SQL Server, and DB2 it became easy to model, store, and transfer data by writing custom stored procedures. Later with the advent of extract, transform, and load tools this process became automated with limited code being written. With the advent of data quality programs more emphasis was given by enterprises to manage the quality of data sourced from source systems and monitored through data governance and data quality programs. Later the concept of information management changed from structured to include unstructured data with the realization that about 80% of all enterprise data are unstructured. This resulted in more disciplines coming into the purview of EIM including ECM. Now with the advent of big data solutions the boundaries of EIM are being expanded further to include unstructured data management solutions as well as new types of data assets such as machine data (sensor data), weblogs, and customer sentiment analysis from social media becoming data sources for EIM. With the increasing amount of use cases for use of diverse data sets to understand enterprise performance, EIM as a discipline is constantly evolving and calling for new types of technologies and processes to analyze and leverage the voluminous data sets currently generated in the digital ecosystem.

CHAPTER 2

■ ■ ■

The Lifecycle of Enterprise Information Management

Enterprises generate data from business processes and tasks. The information generated from analyzing the effectiveness of business processes is the key to managing information in an enterprise. The information that becomes generated has a lifecycle of its own. In this chapter I examine the information lifecycle in some detail—the information lifecycle and business value chain of information. Given is an industry example to demonstrate the business value chain of information.

■ **Note**   Business value chain of information varies widely with industry and hence enterprise information lifecycle management strategies become industry specific. To illustrate, in the insurance industry claims to premium ratio is considered a health indicator for an enterprise. However this is a ratio involving two metrics namely, premium and claims, which in turn can have their own lifecycle and data storage and retention policies. The key is to determine which are the key performance indicators (KPI) of a business and which associated measures or metrics are used in calculating the KPI value. For each of the associated measures, the data lifecycle and retention policies have to consider the business value of the metric. Here business value implies how the metric or measure is used by the business to measure the performance of business processes or enterprise performance.

In any enterprise there are key business processes, such as customer relationship management, supply chain management, knowledge management, operations management, and so forth, which get their information needs from information processing activities such as information integration and storage, transformation, and dissemination. The information process also feeds management activities such as planning, controlling, modelling, and decision making. These in turn give a strategic view of the enterprise market position and its profitability. The metrics that help measure the outcomes and the alignment to business strategy are of higher business value and therefore are considered as moving up the business value chain.

Understanding the Stages of the Life Cycle of Enterprise Information Assets from Creation to Archival

Before embarking on a formal definition of enterprise information lifecycle management (EILM), it is important and noteworthy to understand the lifecycle of enterprise information assets.

Data are created as a part of business processes being executed within and outside an enterprise. The key stages in the lifecycle of information include the steps shown in Figure 2-1. To illustrate with an example, a retail bank has embarked on a customer loyalty program and needs to analyze customer interactions and transactions over a period of the last three years. This means that the transaction data once generated from one of the banking channels needs to be stored in the system of record (SoR) for three years before it can be considered obsolete and marked for archival or destruction. If we look at tax data for individuals and corporations, the government would like to retain this information for at least 20 years and possibly longer. Essentially the nature of the industry, that is, retail banking or government tax departments, defines the retention and archival policies around which data are stored.

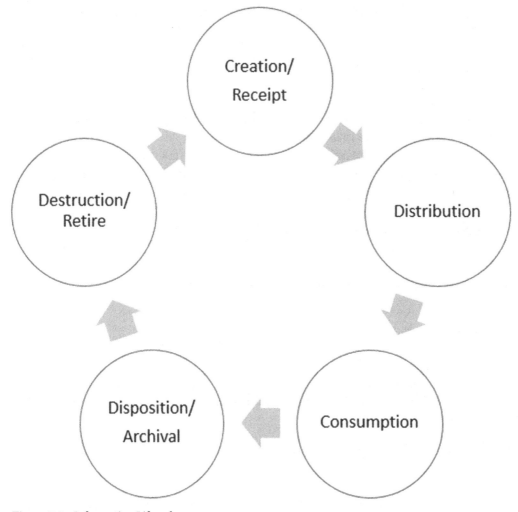

Figure 2-1. Information Lifecycle

The key steps or phases in the information lifecycle are as follows:

- Creation/receipt
- Distribution
- Consumption
- Disposition/archival
- Destruction/retire

Creation/Receipt

Data are created at their point of origin in the business process creating the given piece of information. For instance, when a new customer is acquired, a new customer identifier is assigned and relevant customer attributes are captured. As the customer makes transactions over time, more information is collected and continued right through the lifecycle of the customer. In some cases enterprises depend on external data to augment existing customer data, for example, the DUNS Number for customer matching and demographic information. This is an instance where the enterprise is not creating the data but receiving the data from external data providers. In certain businesses, such as banking and insurance, customer leads also are procured from external customer databases that are based on certain behavioral attributes which are considered suitable for targeting potential customers.

Distribution

Once the data are created, they are distributed to the relevant consumer applications that leverage the data in running core business processes. Let us use this example, once the customer lead is created, the sales department is typically provided the means to create potential new customers by targeting them with specific pitches relevant to their demographic profile and buying patterns. Even for manufacturing companies, once a new product is created the master data are distributed to different departments that need the information. There are multiple ways to distribute the data where the concepts of data integration come into play and will be discussed in Chapter 5.

Consumption

As discussed in the previous step the data are distributed to different consumers, which could be departments, business processes that need data inputs to drive process compliance, as well as completion of transactions. There are principally two types of consumption patterns: 1) end consumers of the data in which the usage pattern is primarily reporting and analyzing (e.g., sales department reporting on leads to customer conversion based on customer data consumed) and 2) business functions and departments consuming data as part of their business process and transaction needs (e.g., actuarial departments needing customer attributes concerning the life style of customers and health parameters while analyzing the risk of a customer profile to issue a life or health insurance policy). Here the data are used for making a decision — whether to issue an insurance policy or not.

Disposition/Archival

The data created and consumed has its own lifecycle that is a key function of the business utility of the data. Each enterprise needs to define the useful period for which a data element needs to be retained in the system of record or other data repositories such as operational data store or data marts. The data lifecycle can be visualized as follows: Data created are treated as active for a period defined by the business where data are either consumed for reporting or analyzing or are used by business processes to complete transactions. After that the data go into a semi-active stage where there are no defined use case for consuming the data element. However the data are still not disposed of as there may be potential use cases in the future, for example, historical analysis of customer loyalty programs or customer behavior patterns. At some point in time, the business would find no case for frequent use of the data and the data element could be marked for archival or disposition. In this step the data elements marked for archival are moved from the system of record or operational data stores to an offline storage mode such as cloud storage in case of hybrid clouds, or optical disk. The focus moving the data to a cheaper means of storage as there is no further use of the data elements in question.

Destruction/Retire

Business processes have their own lifecycle and tend to change over time as businesses change. Due to new stage development, archived data become outdated and no longer relevant to the current business processes and data consumption needs. Data that have reached the end of their lifecycle and thus classified as inactive; the data elements marked as inactive need to be destroyed/retired.

Enterprise Information Lifecycle Management

The entire lifecycle of data elements is called enterprise information lifecycle management (EILM). EILM is the practice of applying policies based on business classification of data for effectively managing information. The basic tenet concerning EILM is to identify and classify data elements based on the business criticality of the data and the nature of the business. The data management policies concerning creation, distribution, consumption, disposition, and destruction then apply to the classified data elements. The Storage Networking Industry Association defines *EILM* as policies, processes, practices, and tools used to align the business value of information with the most appropriate and cost effective IT infrastructure from the time of creation to the disposition phase.

Data classification is part of the EILM process used as a tool for classifying data that can be used by enterprises to answer questions such as: Which types of data are available? Are the data protected with the right controls and do the controls meet the compliance needs as mandated by the industry? Data classification provides the following benefits to enterprises:

- *Data compliance and risk management*

- *Optimization of data encryption needs as all data need not be encrypted*

- *Better control of disaster recovery and business continuity needs*

- *Enhanced metadata management as result of classification of data assets*

- *Appropriate data security controls and access based on the criticality of the data*

EILM is fairly new to enterprises and adoption trends are still around 30 to 40%. However with government laws, such as the Health Insurance Portability and Accountability Act (HIPAA) that have varying directives concerning the length of time patient records need to be retained, there now is a pronounced need for organizations to have an EILM strategy as part of their information management strategy. The EILM strategy comprises of the following steps:

- *Data classifications of data assets based on business value*

- *Assess data security needs based on data classifications*

- *Define and implement policies concerning data type and storage mechanisms*

The EILM implementation in any enterprise consists of the following steps (Figure 2-2):

- *Analyze/categorize—classify data assets based on business value and technical usage patterns (query usage, usage statistics, etc.)*

- *Develop strategy—develop data aging and data retention strategy based on business needs and data classification. Develop data security needs based on data classification.*

- *Implement strategy—implement data aging and data retention strategy as well as data security needs. Schedule monitoring and archiving processes.*

- *Monitor—monitor data archiving and data retention processes.*

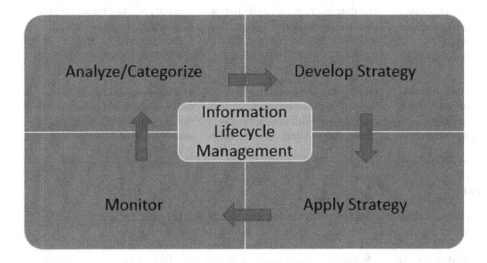

Figure 2-2. *EILM road map*

Understanding the Actors in the Stages of the Enterprise Information Lifecycle

In this section we look at each of the stages of the enterprise information lifecycle and the associated actors in these stages. This understanding is crucial for enterprises to have authority concerning the usage and classification of data assets as well as agreement between business and IT concerning data retention and archiving principles. Table 2-1 maps the stages of the enterprise information lifecycle to the associated actors. For details concerning the responsibilities associated with the actors see the descriptions after Table 2-1.

Table 2-1. *Stakeholder Mapping to Each Stage of Enterprise Information Lifecycle*

Information Stage	Actors	Constraints	Comments
Creation	Business process, business users, data stewards	Only the relevant users need to have access to create the required data assets.	Creation of master data needs to be handled by the business process and user with create rights for the given master data.
Distribution	Data integration processes, data integration teams, data stewards	The data distribution/synchronizations processes would run on a scheduled or on a demand basis.	The data distribution/synchronization processes would feed the downstream applications/business processes that need the data for reporting and transaction processing needs.
Consumption	Consuming business processes and business users	The consuming applications/business processes are provided the relevant data sets based on their needs with data security controls in place.	Data would be consumed in with modes as flat files for further processing or as access to data repositories such as system of record (SoR), data marts, and so forth.
Disposition	Data stewards, infrastructure architect, business users	The data disposition would happen once the data became semi-active (business defined) and the data are marked for moving to offline storage.	The data marked for archival/disposition would be moved to cloud storage/optical disks in consultation with the infrastructure architect.
Destruction/retire	Data stewards, infrastructure architect, business users	The destruction/retiring will happen once the business in consultation with data stewards agrees that the data can be marked as inactive.	The data marked for destruction would be moved out of optical disks/tape in consultation with the infrastructure team.

Here is the list of actors:

- *Business users*—business users define the business processes that generate the transaction or master data. They also define the key data attributes of a transaction based on the business process. Business users also review and approve the master data created by data stewards. Business users also define when the data are no longer considered active and can be marked for disposition or retiral.

- *Data stewards*—data stewards are data custodians of the enterprise data assets created as part of business process execution. They control access to data assets based on user role and the prevailing data management policies. They also define the data management policies in consultation with the data governance organization in place.

- *Data integration teams*—data integration teams define the data synchronization/data replication jobs that are needed to distribute data needed by consuming business processes and applications. They work closely with data stewards to understand the data consumption needs and the relevant data sets that need to be distributed.

- *Infrastructure architect*—infrastructure architect defines which offline storage media would be used for storing semi-active data. They also work closely with data stewards and business users to identify which data assets need to be moved to offline storage as well as which are to be marked for retirement/destruction.

With a better understanding of each of the stages in the enterprise information lifecycle and the associated actors, in the following, I describe the roles of each of these actors to help you understand their involvement at each phase of the information lifecycle. I also look at the organization model (representative) to support the information lifecycle.

Information Lifecycle - Actors and Their Roles

Table 2-2 provides a breakdown of the actors that are involved in the information lifecycle and their roles.

Table 2-2. *Actors Involved in the Information Lifecycle and Their Associated Roles*

Information Stage	Actors	Roles
Creation	Business users and associated business process	Creation—business users related to business processes are closely involved in the creation/approval of master data/reference and transaction data.
Consumption	Business user and associated business process	Consumption—business users can either consume the transaction and master data through or use the data for further transaction processing or for approval of transactions.
Disposition	Business user and associated business process	Disposition—business users associated with a business process that either create or consume the data need to provide their approval to data stewards concerning marking the aged data as semi-active where the data can be moved to offline storage media.
Destruction/retire	Business user and associated business process	Destruction—business users associated with the business process that consumes the semi-active data elements in question need to provide their approvals to data stewards to ensure that the semi-active data can be marked as inactive with no further business use.
Creation	Data stewards	Creation—data stewards facilitate the creation of master and reference data and obtain the required approvals from business users. While transaction data are created by business processes that manage transactions, data stewards are involved in the validation of transactions with business users.
Distribution	Data stewards	Distribution—data stewards work closely with data integration teams to design which business processes and users have access to which data.
Disposition	Data stewards	Disposition—data stewards work closely with business users who either create or consume the active data to determine whether the data can be marked as semi-active based on aging and business needs.
Destruction/retire	Data stewards	Destruction—data stewards work closely with business users who consumed the semi-active data to determine whether there is any further use of the data anticipated and whether the data can be marked as inactive.

(continued)

Table 2-2. (*continued*)

Information Stage	Actors	Roles
Data distribution	Data integration team	Data distribution—the data integration team designs the data integration/synchronization jobs that distribute the data to consuming business processes/applications. They work closely with data stewards to define who has access to which data.
Data disposition	Infrastructure architect	Data disposition—the infrastructure architect works closely with the data steward to understand which data have been marked as semi-active and can be moved to offline storage media.
Data destruction/ retire	Infrastructure architect	Data destruction/retire—the infrastructure architect works closely with the data steward to understand which data have been marked as inactive and can be destroyed/retired.

Information Lifecycle—Organization Model

As I analyzed the different stages in the information lifecycle and the associated actors, it is also crucial to understand the organization model that supports and enables the information lifecycle. Figure 2-3 represents the enterprise information lifecycle phases mapped to the actors in the organization model.

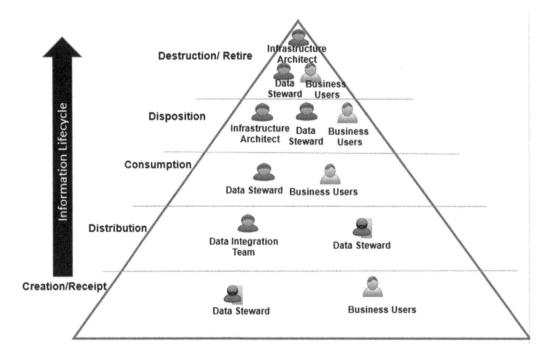

Figure 2-3. *Information lifecycle organization model*

The organization model is to support the information model from the point of creation/receipt to the end-of-life/destruction stage. At each stage different actors support and manage the information assets based on policies defined in consultation with business.

CHAPTER 3

■ ■ ■

Components of Enterprise Information Management

With an understanding of what enterprise information management (EIM) is and how the information lifecycle works from creation/receipt to retirement, I now move to the core components of an EIM solution. The key or core components of an EIM solution are associated around the governance; quality of the data sourced for consumption; the integration patterns of data types; the associated data architecture and data models; the ability to manage associated master data and metadata; and the challenges posed by the increasing interest in new data types such as machine data, sensor data, weblogs, and call center records. The key solution components of an EIM solution are as follows:

- *Information sourcing*

- *Information integration and exchange*

- *Information governance and quality*

- *Information architecture and models*

- *Master information management*

- *Information warehousing and reservoirs*

- *Information delivery and consumption*

- *Metadata management*

- *Big data components*

I cover each of these components in detail in the following sections and later in greater detail in the following chapters (Chapters 4 through 11). Before getting into the details of each of the solution components of an EIM solution, it is better to have a bigger picture or architecture blueprint for EIM in a given enterprise. This serves as a reference architecture to which the business needs of the enterprise can be based on and can be addressed in an incremental fashion to provide business benefits. The next section discusses EIM reference architecture and covers the solution components mentioned.

■ **Note**   The focus of this chapter is the components of enterprise information management from a structured data management perspective. However it is important to note that all of these concepts and frameworks can be applied equally to unstructured and semi-structured data. In this age of big data applications the lines between structured and unstructured data are getting blurred with end consumers of data more interested in the delivery and dissemination of data rather than the source of the data in question. The components of an EIM solution ensure that enterprise have the suitable information to address business questions concerning the performance of the enterprise. Trusted information is available at the right latency, to the appropriate consumers with the suitable granularity.

Enterprise Information Management—Reference Architecture

Before embarking on an EIM implementation, it is essential for enterprises to have a reference architecture that would serve as a blueprint for all EIM solutions. It is imperative to understand that the architecture constructs can be built over time and the reference architecture is designed to keep in mind the business strategies and priorities of the enterprise.

Let's take an example to illustrate this point—for a large retail bank that needed the information latency of minutes to respond to customers, they felt the need for real-time data integration. This business driver was translated to a need for change data capture (CDC; an integration style) that needs a data integration capability which can track changes in source data (transactions) in near real time. The reference architecture for such a bank would entail having CDC as an integration pattern with CDC agents installed in source system applications or a replica of these applications to scrape the transaction logs and analyze the changes and then communicate these changes through the integration platform into the key decision support repositories such as data warehouses and operational data stores (ODS). In some industries where collaboration extends beyond the enterprise boundaries into suppliers and vendors; the need for information exchange between these parties provides new challenges, which needs to be addressed through the reference architecture.

Now let's look at a standard reference architecture for EIM (see Figure 3-1, for an industry agnostic view). The information is sourced from a series of business applications such as custom applications, package applications, manufacturing systems, and external data sources. The information sourced is then integrated using a series of integration styles such as ETL/ELT/SOA and loaded into a set of repositories such as master data stores, ODS, data warehouses (system of record; SoR), data marts, reference data stores, enterprise content management repositories, and so forth. The information sourcing has accepted protocols concerning information exchange standards, governance procedures, and associated data stewardship services.

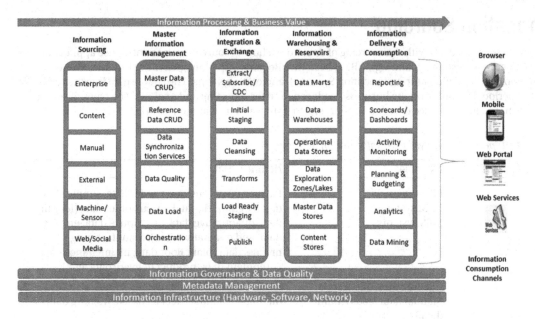

Figure 3-1. *Enterprise information management—Reference architecture*

The key drivers for a reference architecture for EIM are as follows:

- *Data integration and single version of truth*—Data is integrated across multiple subject areas (e.g., customer, product, geography, employee, etc.) to provide a single version of truth for consistent decision making.

- *Real-time business agility*—Continuous and seamless information across transactional, operational, and analytical environments to support real-time business agility.

- *Incremental business value*—Component-based solution components to address business priorities that provide business value on an incremental basis.

- *Business productivity gains*—reduced time to discover, analyze, use, and act on information due to more consistent data definitions and data services.

- *Data governance*—data services to govern, move, cleanse, profile, transform, access, and enrich information.

- *Metadata management*—metadata management and semantic reconciliation to ensure consistency, encapsulation, transparency, and integrity of information assets.

The reference architecture is comprised of key layers or components that deal with the information lifecycle from sourcing/creation to delivery and consumption. The layers can be built in an incremental manner and the architecture evolves with time as the organization matures.

The next sections cover each layer or solution component of the EIM reference architecture.

Information Sourcing

Information sourcing deals with the sourcing of data from a host of sources such as enterprise applications as ERP (Enterprise Resource Planning), CRM (Customer Relationship Management), SFA (Sales Force Automation), MES (Manufacturing Execution Systems), and other systems. The source data is extracted on a defined period depending on the business needs required for reporting and analysis. The types of sources that contribute source data for decision support systems such as data warehouses are shown in Table 3-1.

Table 3-1. *Types of Source Systems—Information Sourcing*

Type of Source Data	Examples	Nature of Data
Enterprise	ERP systems CRM systems MES systems	Payment data, customer order data, supply chain data, organizational data, customer data (identity data, quantitative data, descriptive data, employee data, and qualitative data). Manufacturing process data (plant floor data, maintenance data, equipment data, product delivery data, etc.)
Manual	Classification data excels Manual data feeds (training details, etc.)	Classification data maintained by business manually. Manual excel feeds like training data.
External	Dun and Bradstreet (D&B) Number IMS data	Customer demographic information purchased from external data providers such as D&B. Sales data, prescription data, and medical claims data from providers such as IMS.
Machine/sensor	Web server logs Call detail records Smart electric meters Engine data	Web server: client IP address, request data, HTTP code, bytes served, referrer data; Call detail records: calling party data, receiving party data, call duration, billing number, routing data, etc.; Smart meters data: meter deployment data, network asset, monitoring data, electricity consumption data, etc.; Engine data: engine model, thrust, flat rating, bypass ratio, pressure ration, etc.
Social media	Social media network Social search Publishing platforms Content sharing platforms	Social media network data: tweets, posts, sentiment data; Social search data: keywords analysis and hashtag data; Publishing platforms: blogs, wikis, opinion sites, etc.; Content sharing platforms: SlideShare, YouTube, etc.

With an understanding of the different types of sources, it also is important to understand the extract mechanism for the source data to be delivered for absorption by the information integration mechanism. There are primarily two ways in which source data extraction can happen. Table 3-2 illustrates the key differences between push and pull mechanism.

- Push mechanism—This includes the source systems teams generating source data extracts that are they moved through file transfer mechanisms to a file landing zone from which the files are picked up for processing by the information integration engine. The key point here is to use full source extracts for master or dimension data and incremental extracts for transactional data (where only delta changes are marked for consumption). In the push mechanism, the information integration engine team publishes the interface data needs and the source system team codes source extracts based on the interface agreement with the information integration team.

- Pull mechanism—This involves the information integration engine to access the source system database, query, and pick up the relevant data sets from the source system. Here the onus is on the information integration engine to pull the right data sets for further processing.

Table 3-2. *Key Differences Between Push and Pull Mechanisms*

Parameters	Push Mechanism	Pull Mechanism
Nature of extraction	Source system team provides the source data extracts in the interface formats provided by the information integration team.	The information integration team is provided read access to source tables to query and pick up the relevant data sets for further processing.
Source system knowledge	The source system team has extensive knowledge of the source systems and provides the source data extracts as per the interface formats agreed with the information integration team.	The information integration team has to build knowledge of the source system and **extract** the relevant data from the source tables based on the access provided by the source systems team.
Source system changes	In case of push mechanism, there is no impact of source system structure changes as the source system team generates the extract files. The information integration process is insulated from the source system changes.	In case of pull mechanism, the source system structure changes have to be understood by the information integration team and there will be changes to the information integration jobs that access the source systems to pull relevant data.

Information Integration and Exchange

Information integration and exchange is the process by which the information sourced is ingested by the information integration engine. In cases of information exchange with external systems or other internal systems that consume the data generated by the information integration process, there needs to be an interface agreement about the data which are exchanged between the consuming system (subscriber) and publishing system. There are numerous integration styles, the key ones include the following:

- Full extraction—Full extraction is used when the data need to be extracted and loaded for the first time (e.g., history data load). In full extraction, data from the relevant source are extracted completely. These data reflect the current data available in the source system.

- Incremental extraction—In incremental extraction, only the changes in source data need to be tracked since the last successful extraction. Only the changes in the data will be extracted and subsequently loaded. These changes can be detected from the source data that have the last changed time stamp. In addition a change table can be created in the source system, which keeps track of the changes in the source data.

- Change data capture—CDC is the process of capturing changes made at the source system and applying these changes throughout the enterprise for both decision support systems, such as operational data store or data warehouses, as well as other downstream consuming applications. CDC reduces the ETL resources needed as it only picks up the changes in the source data. There are multiple ways in which CDC can be tracked.

- Slowly changing dimensions (SCD)—SCD are dimensions that change slowly over time. Often in data warehouses and data marts there is a need to track changes in dimension attributes to report historical data trends. There are multiple approaches to SCD including Type 0, Type 1, Type 2, Type 3, and Type 6. Type 0 is the passive method where no changes are made even when the dimension attribute changes. In Type 1 SCD the older attributes values are overwritten with the new value (hence no history is maintained). In Type 2, each time the dimension attribute value changes a new record is inserted whereby complete history is maintained. In Type 3, a new column is inserted where the current value and previous value of the dimension attribute is captured (hence only partial history is retained). In Type 4, a separate historical table retains the historical values of the attributes while the main dimension table only retains current value of the attribute. In Type 6 there is a combination of Types 1, 2, and 3.

- Data integration hubs—A common integration pattern in which all data integration needs are handled by a data integration hub and all-consuming applications are like spokes emerging from the hub (the hub and spoke model). This ensures that the principle of extract once and consume many times is followed at all times. The hub has the extracted data as part of the integration layer which can be consumed by the spokes multiple times.

- Outbound extracts—In many instances information is exchanged between enterprises and partners and this is through a set of agreed data formats in the form of outbound extracts files. The data can be generated in CSV/XML formats for consumption by external systems.

Information integration and exchange is covered in greater detail in Chapter 5.

Information Governance and Quality

Information governance and quality is one of the pillars of EIM. Although a lot of enterprises look at data quality from a tactical standpoint and tend to pay lip service to information governance, it is a crucial part of managing enterprise data. Information governance ensures that enterprise data is trusted and its usage is governed based on the classification of the data being consumed and the rights of the requestor. Information governance is not about technology alone but about people taking responsibility for the information assets of their organization by looking at the processes they use to interact with information as well as how and why it is being used.

Creating a governance framework to ensure the confidentiality, quality, and integrity of data—the key drivers of information governance is essential to meet both internal and external requirements such as financial reporting, regulatory compliance, data security, and privacy needs. Information governance manages risk, both operational and compliance risk by increasing oversight. Information governance enables enterprises to integrate and consolidate information from vertical and horizontal lines of business into a single version of truth, providing a bird's eye view to enterprise performance and helps tie information policies to business strategy.

Information quality is one of the enabling pillars of information governance. Although there is a debate concerning what comes first, organizations must look at information quality as one of the enablers of information governance. As shown in the reference architecture, information governance and quality cuts across each stage of information processing from information sourcing to information delivery and consumption. Although it is more challenging to define a business case for information governance, it is easier to do so for information quality as it can be tied to direct business benefits such as reducing duplicates in the customer master results in percentage of dollar savings in a specific campaign management exercise or in direct marketing costs. Information governance and quality is covered in greater detail in Chapter 6.

Information Architecture and Models

Information architecture and models deal with the how the information flow is designed to support information delivery and consumption as well as promote efficiency in information processing. Information architecture defines the blueprint for information modeled to support business information and analytical needs. Information models deal with how data are modeled to support decision support needs and analytical needs. There are numerous ways in which information models can be designed to support the business requirements for reporting and analysis. There are mainly three approaches that are followed in the industry namely:

- Top-down approach—with this approach the enterprise business requirements are modelled as a set of entities and relationships between the entities (also known as entity relation modelling or ER modelling). The SoR (EDW) is designed as an ER model. It serves as the repository for all enterprise data. For any specific function and departmental reporting needs data marts are designed from the data warehouse. The data marts can be seen as dimensional models with a set of facts and dimensions. The dimensional model is catered around the reporting requirements and ensures optimal query performance. Most of the reporting happens out of the data marts that are a subset of the EDW. This approach has been defined by the father of data warehousing, Bill Inmon.

- Bottom-up approach—with this approach business reporting requirements are modelled as a set of facts and dimensions into a set of star schemas. The conformance of common dimensions (enterprise dimensions) is enforced across the star schemas. This is often known as a dimensional modelling approach to data warehousing. This approach was defined by the father of dimensional modelling, Ralph Kimball.

- Data vault modelling—a modelling construct that covers three main components, hub, link, and satellite. The hub represents a key business concept such as customer, supplier, sales, or product. Links represent natural business relationships between business keys, for example the hubs for customer and product through a sale transaction. The descriptive attributes and temporal attributes of hubs and links are stored in the satellite tables. Dan Linstedt is considered the founder of this modelling approach.

Information models usually start at the conceptual level with a business understanding, which is then translated into a logical level with the capture of relationships between the entities or facts and dimensions. Finally the physical model is generated, which looks at specific nuances concerning the database platform that would host the data warehouse or data mart in question.

Master Information Management

Master information management deals with the business processes, governance policies, standards, and technologies that consistently define and manage the critical master data entities such as customer, product, vendor of an organization, and provides a single version of truth. Master data management ensures that an organization does not use multiple versions of the same master data in different parts of its business operations, resulting in higher costs. Some of the common issues with master data management include inconsistent data quality, ownership issues, classification and identification of reference data, and data reconciliation issues as different versions of master data reside in multiple source systems. Master data management of disparate systems involves data transformations as the data extracted from diverse source systems is transformed and loaded into a master data repository (hub). Once the master data resides in the master data repository it needs to be distributed to other consuming applications.

In the age of big data solutions, master data management remains a key component for addressing challenges around customer centric objectives. To improve customer experience, organizations need to do what master data has been doing for long time—build a golden record for customer masters. There are four key implementation styles for master data implementations. They are as follows:

- Registry—With this style, the system of record for the master is the source systems. Therefore changes to master data continue to be made through the existing source systems. Only enough information to match and link between similar or matching records is stored and a trusted view of this information is provided to the end users.

- Consolidation—This style matches and physically stores a consolidated view of master data in the central hub. The authoring of the data remains distributed across the spoke systems and the master data are updated after events and are not guaranteed to be up to date. There also is no real-time publish and subscribe. The master data in this case usually are not used for transactions, but rather support reporting; however, the data also can be used for reference operationally.

- Coexistence—Coexistence style MDM hub involves matching and physically storing a consolidated view of master data in the hub; publishing a consolidated view of master data for harmonization across systems and for central reference. The data are updated after the event and are not up to date.

- Centralized—The centralized or transactional style is an approach in which all information relevant to providing a single view is loaded from the source systems into a new central repository. The matching and collapsing of records across the source systems are done and a single view of the party is typically persisted and updated in this repository. The MDM system now becomes the SoR for storing master data.

Information Warehousing and Reservoirs

Information warehousing and reservoirs is an evolving concept as the information management industry matures toward new types of data and technologies to process them. Information warehousing deals with all the data repositories in the reference architecture including the following:

- Data warehouses—the system of record or enterprise data warehouse repository

- Data marts—specific to the function (e.g., sales data mart repository) or departmental data mart repository (e.g., finance data mart). The data can be retained for multiple years for business users to perform time series analysis.

- Operational data stores—ODS serve as repository for all operational data needs. The data are usually retained for a period of three to six months as operational data needs are usually recent data.

- Master data stores—serve as a repository for key master data entities such as product, customer, suppliers, and so forth. Enterprises also invest in reference data stores for storing commonly used reference data such as currency codes, country codes, and so forth.

- Content stores—serve as a repository for all enterprise content such as contractual documents, policy documents, e-mails, and so forth. Content stores serve as a repository for unstructured content.

- Data lakes/data exploration zones—a relatively new concept with the advent of big data solutions, data lakes are a collection of repositories for both structured and unstructured data and serve as data exploration zones for running algorithms and data visualization for insights into enterprise data.

Information warehousing and reservoirs are covered in greater detail in Chapter 8.

Information Delivery and Consumption

Information delivery and consumption deals with a set of information delivery approaches and consumption styles through which the information processed from sourcing to various repositories can be consumed. The key delivery approaches in use are as follows:

- Reports—can be operational, canned, scheduled, or ad hoc in nature. Reports can be delivered through e-mail, portals, or through bursting mechanisms.

- Scorecards/dashboards—can be operational, tactical as well as strategic depending on the nature of analysis and end user base. Enterprises often use balanced scorecards and dashboards to monitor execution of strategy by measuring key result areas through a set of business critical KPIs. There could be lagging and leading indicators of business performance.

- Activity monitoring—Business activity monitoring involves providing near real-time visibility into business activity and processes. This helps manage service level agreements with processes as well as providing visibility into process performance through dashboards.

- Planning and budgeting—Planning and budgeting applications also consume data (like plan data) from the data warehouse and data marts. Outbound feeds serve as data inputs into planning and budgeting applications.

- Analytics—Analytical tools often get data feeds from data warehouses and data marts. They also can read data from data marts as input for running analytical models for training the data and find insights.

- Data mining—Data mining tools also often get feeds from data warehouses, ODS, or data marts depending on the nature of business question needed to answer.

Metadata Management

Metadata management is another key pillar of EIM. *Metadata* is often defined as data about data and provides a context to the data it is associated with. Metadata can be managed through a set of defined processes wherein metadata is captured at each stage of a data management project. This ensures complete data lineage and traceability of data attributes as they move from information sourcing to information delivery and consumption. The common metadata types include the following:

- Business metadata— includes business requirements, business metrics, and key performance indicators, business terms, business process flows, and so forth.

- Technical metadata—captures details concerning data models; ETL job designs; and ETL mapping metadata, which provides detailed insights into the data processing job design and workflows.

- Operational metadata—captures details concerning operational metrics of running the ETL batch or reports batch. Statistics collected can be analyzed over time to monitor trends as well as look at data quality issues.

Details about metadata process and tools are covered in Chapter 10.

Big Data Components

Big data is a new buzzword that is taking the information management world by storm. However it is important to understand that big data solutions augment the functionality provided by traditional information management solution components. The key big data solution components are as follows:

- Information sourcing—Big data involves a new variety of data sources and huge volumes of data that need to be sourced. Some big data sources include weblogs (from e-commerce sites) and social media feeds from Twitter and Facebook. Machine data from sensors on an oil rig or manufacturing plant.

- Information integration and exchange—These new data types need new styles of ingestion and integration. Big data can be integrated using ETL tools with support for big data such as Informatica, DataStage, or Talend or even with tools that work within the Hadoop ecosystem such as Apache Hive or Apache Pig. Hive and Pig are good at loading unstructured, structured, and semi-structured data into the Hadoop distributed file system (HDFS).

- Information reservoirs/lakes—The concept of information reservoirs and lakes has emerged with the rise of semi-structured and unstructured data in the big data landscape. These include both structured repositories such as traditional data warehouses and data marts and repositories such as HDFS to store unstructured and semi-structured data.

- Information visualization—As the information lakes have both structured and semi-structured data there is a need for visualization tools that can search and present to users a unified view of both structured and unstructured data.

The big data solution components that enhance the EIM capabilities of an organization are covered in greater detail in Chapter 11.

CHAPTER 4

■ ■ ■

Pillar No. 1: Information Sourcing

With an understanding of enterprise information management (EIM) and how the information lifecycle works from creation/receipt to retirement, and the core components of an EIM solution, we now move to information sourcing. Information sourcing deals with the sourcing of source data from a host of sources such as enterprise applications like ERP, CRM, SFA, MES, and other internal and external systems. Information sourcing is the first step in the EIM reference architecture and the starting point for any data warehouse program or analytics initiative. Once the business requirements have been documented, information integration teams start talking to the data sourcing subject matter experts (SMEs) to understand which data resides in the source systems and the best way to extract the data. The source data is extracted on defined periods depending on the business analytics needs. Some of the key steps followed in data sourcing are as follows:

- *Mapping business requirements to source systems*

- *Profile source systems for relevant data sets*

- *Define source extraction mechanism*

- *Provide source extract files for information integration*

In the next section I discuss the different sources where organizations can source their information.

■ **Note**   To explain the different types of sources where information is sourced, the different approaches to information sourcing with relative cost benefit analysis, you need to understand the patterns of information sourcing, their challenges and how these challenges can be addressed. You also need to understand how the information is generated and the importance of granularity in the information sourcing process.

Information Sourcing—Types of Sources

As mentioned in Chapter 3 there are a multitude of sources. The common types of sources include enterprise data sources, such as enterprise resource planning systems (e.g., SAP, Oracle, etc.); customer relationship management systems (e.g., SAP, Pivotal, Microsoft Dynamics, etc.); manufacturing executing systems (e.g., GE, Honeywell, Schneider Electric, etc.); manual data sources, such as classification data, reference data, external data (e.g., customer identification numbers like DUNS or sales data from IMS, AC Nielsen, machine/sensor data from smart energy meters, and social media data from websites like Twitter, YouTube, etc.). See Chapter 3 Table 3-1 for types of source system data that are involved in information sourcing.

With an understanding of the different types of sources it also is important to understand the key steps in information sourcing, which are covered in following sections.

Mapping Business Requirements to Source Systems

One of the key steps in an EIM program is to ensure that business requirements are mapped to source systems. The business requirements can be broken down into key performance areas that are measured through key performance indicators (KPIs) and measures. The KPIs can be presented through a KPI dimension matrix (see Figure 4-1) and the KPIs and measures need to be mapped back to the requisite source systems. The KPI dimension matrix is a vital tool for translating business requirements into the data strategy for the successful execution of an EIM program. The KPI dimension matrix (see sample KPI dimension matrix in Figure 4-1) also plays an information role in defining the granularity at which the KPI or measure is monitored that has a bearing on the data sourcing strategy as well as the frequency of the data extraction mechanism.

		KPI Dimension Matrix							
Business Function ▼	Functional Area ▼	Key Performance Indicator ▼	Granularity ▼	Base/Derived ▼	Employee ▼	Time ▼	Business Unit ▼	Dimension 4 ▼	Dimension 5 ▼
HR	Employee Adminstration	Employee Engagement Score	Quarterly	Derived	X	X	X		
HR	Sourcing	Time to Hire Critical Positions	Quarterly	Derived	X	X	X		
HR	Retention	Employee Attrition %	Monthly/Quarterly/Yearly	Derived	X	X	X		

Figure 4-1. *A sample KPI dimension matrix*

Once the KPI dimension matrix is defined the data sourcing team talks with the source system SMEs in mapping the measures to the source systems.

Profile Source Systems for Relevant Datasets

Once the measures are mapped to source systems, the next step is to identify the right source of data for specific measures and get down to column level mapping between the measures and source systems. Often at this stage the data sourcing team is met with a dilemma, the possibility of mapping the same measure to multiple columns in from different source systems. Although in some organizations there is subject matter expertise to resolve such mapping issues, it is a recommended practice to profile data sources before such mapping decisions are made. Data profiling helps not only in understanding the underlying data quality of the source data in terms of accuracy, recency, and reliability but also helps to validate the metadata associated with the source data. Data profiling also can be used to discover metadata where none exists about source systems data. The metadata about the source data that is discovered during the data profiling exercise helps to determine illegal values, duplicates in source data, data in incorrect formats, and other data quality issues. Once the profiling data is examined in consultation with the source system SMEs, decisions can be taken with a higher degree of confidence concerning mapping of the source system columns to the measures. Although data profiling can be performed multiple times in the lifecycle of an EIM program, the first instance of profiling should be done as part of source system mapping to find out the relevant data sets from a data sourcing perspective.

Define Source Extract Mechanisms

Once the source data has been profiled, the results of the profiling analyzed, and the mapping to source systems done, it is time to define the source system extract mechanisms. As explained in Chapter 3, there are primarily two extract mechanisms in use—push and pull mechanisms. Normally it is recommended to go with the push mechanisms where the source system team generates the source extract data files based on the interface agreement defined by the information integration team. The benefit of this approach is that source system changes do not impact the existing extracts unless they need to add additional columns based on the source system changes. The different approaches are covered in the following sections.

Provide Source Extract Files for Information Integration

The source system teams define the source data extracts based on the interface agreement defined by the information integration team. The interface agreement has the following details:

- Structure of data file—with all columns, data types, and precision needs (e.g., floating points with number of decimals defined)

- Naming convention of data files and header files with details such as subject area, extraction time stamp, and so forth.

- Associated header file with details concerning number of records, checksum values for a couple of key measures in the source data files.

- The frequency of the generation of data files (such as daily, weekly, on demand, monthly, etc.).

- The mode of delivery of the data files—file transfer method to a specific folder as file landing area.

Information Sourcing—The Different Approaches

Information sourcing deals with the sourcing of source data from a host of sources such as enterprise applications like ERP, CRM, SFA, MES, and other systems. The source data is extracted from a defined period depending on the reporting frequency of the measures that are being monitored and reported. There are different approaches to information sourcing. The key differences between push and pull mechanism are shown in Table 4-1.

- Push mechanism—This includes the source systems teams generating source data extracts that are moved through the file transfer mechanisms to a file landing zone, where the files are picked up for processing by the information integration engine. The key point here is to use full source extracts for master or dimension data and incremental extracts for transactional data (where only delta changes are marked for consumption). In the push mechanism, the information integration engine team publishes the interface data needs (interface agreements) and the source system team codes source extracts based on the interface agreement with the information integration team.

- Pull mechanism—In this mechanism, the information integration engine is provided the required access to the relevant source system database tables and acquire the relevant data sets from the source system. Here the onus is on the information integration engine to pull the right data sets for further processing. The source system team's involvement here is to provide mapping of the measures to the source system columns and provide read access to the source system tables for the information integration team to execute the data sourcing queries and pull the required data.

Table 4-1. *Key Differences Between Push and Pull Mechanism*

Parameters	Push Mechanism	Pull Mechanism
Nature of extraction	Source system team provides the source data extracts in the interface formats provided by the information integration team.	The information integration team is provided read access to source tables to query and pick up the relevant data sets for further processing.
Source system knowledge	The source system team has extensive knowledge of the source systems and provides the source data extracts as per the interface formats agreed with the information integration team.	The information integration team has to build knowledge of the source system and extract the relevant data from the source tables based on the access provided by the source systems team.
Source system changes	In the case of push mechanism, there is no impact of source system structure changes as the source system team generates the extract files. The information integration process is insulated from the source system changes.	In the case of pull mechanism, the source system structure changes have to be understood by the information integration team and there will be changes to the information integration jobs that access the source systems to pull relevant data.
Performance of Source Systems	As the source system teams provide the source data extracts as per the defined schedule, there is no impact on the source systems due to the information integration processing of source data.	As the information integration process has to access source system tables and query them to pull the source data there is bound to be some performance impact on the source system performance.
Source System outages	Push mechanisms are not impacted by planned source system outages as the source extracts can be generated before the system downtime.	Source system outages result in the pull mechanism not working as the integration process cannot execute the queries to pull the source data from source systems.

As is evident from the comparison of push and pull mechanisms, it is clear that push mechanisms are a better approach from a reusability and performance standpoint. In the case of pull mechanisms the extraction logic has to be written by the integration team who may have limited understanding of the source systems, and it creates opportunities for incorrect extraction logic or lack of knowledge of source system changes over time resulting in the source data extraction logic to be rewritten numerous times. Also from a security standpoint, source systems have critical mission information concerning business processes and providing access to source system tables can cause data security issues.

Information Sourcing Patterns and Challenges

Information sourcing is the process by which the information sourced is absorbed by the information integration engine. In cases of information exchange with external systems or other internal systems that consume the data generated by the information integration process, there needs to be an interface agreement based on which data is exchanged between the consuming system (subscriber) and publishing system. There are different information sourcing styles or patterns by which data is extracted from source systems. They are as follows:

- Full extraction—Full extraction is used when the data needs to be extracted and loaded for the first time (e.g., history data load). In full extraction, data from the relevant source is extracted completely. These data reflect the current and historical data available in the source system. In cases of dimensions or master data that do not undergo too many changes over time a full extract may suffice from a data extraction strategy standpoint. In full extracts there is no need to track changes in the data source since the last extraction. Hence full extraction cannot be used as a data extraction strategy for slow changing dimensions where a history of changes needs to be retained from a reporting and analytics standpoint.

- Incremental extraction—In incremental extraction, only the changes in source data need to be tracked since the last successful extraction. Only the changes in the data will be extracted and subsequently loaded. These changes can be detected from the source data that have the last changed time stamp. In addition a change table can be created in the source system, which keeps track of the changes in the source data. There also can be an extraction based on the time stamp where only the source records inserted/updated since the last extraction will be picked up from the source systems.

- Change data capture (CDC)—CDC is the process of capturing changes made at the source system and applying these changes throughout the enterprise for both decision support systems such as operational data store or data warehouses as well as other downstream consuming applications. CDC reduces the ETL resources needed as it only picks up the changes in the source data. There are multiple ways in which CDC can be tracked. CDC enables incremental data delivery that is used for 1) publishing to consuming systems and 2) real-time integration. It involves a log-based capture of database changes with minimal impact on source systems. Commercial CDC packages support a wide variety of sources and targets. Some typical use cases of CDC include a) application consolidation and migration, b) ETL batch window reduction, and c) data synchronization and live reporting by moving operational data to a secondary system such as ODS or data warehouse.

The other data integration patterns are more centered on operational data stores or data warehouses and are covered in Chapter 5. Some of the key challenges concerning data sourcing are as follows:

- Data conversion challenges—Often data has to be sourced from legacy systems that may have different granularity and coding mechanisms for enterprise data, such as employee codes, and may require significant mapping revisions to achieve an acceptable outcome. In case of granularity mismatch between the data source and the data needed in the downstream systems the issue is more complex and often needs business intervention in building transformation rules where data sanctity may be lost to some extent. This can often lead to project delays.

- Metadata gaps—lack of metadata concerning legacy systems results in further challenges in data conversions. This can cause serious issues in data quality as partially understood metadata often leads to data conversion errors.

- Mergers and acquisitions—makes data sourcing a very challenging exercise as master data entities are defined differently. Transaction granularities may mismatch resulting in serious data sourcing challenges. Even the data quality checks and balances may vary widely from one enterprise to another.

- Manual data—Despite numerous automated cycles in today's enterprises, there are still silos of information maintained in spreadsheets (e.g., planning data) that need to be integrated into the data warehouse or data marts. Although there may be some validation built in the spreadsheets, there are significant chances of typing errors and data quality errors being introduced while managing these spreadsheets. Another issue could be that users assign dummy values where proper classification codes exist and have not been mapped.

- Real-time source data extraction—Real-time interfaces propagates data quickly to a downstream consuming system, but they also provide limited capabilities to check the quality of data loaded to the downstream systems. Real-time data quality checks are more complex to implement and do bring some latency to the data availability needs.

Given these challenges it is always imperative to provide time for source data analysis before the integration strategy and design is performed. Often source data analysis issues are captured later in the data integration lifecycle causing significant cost and project overruns to integration projects.

Information Sourcing—The Importance of Granularity

Information granularity refers to the level of detail available in the source data elements. The more detailed data that is available in source systems, the lower the level of granularity. Likewise, the less detailed data in the source systems implies a higher level of granularity. Granularity of data in the system of record or data warehouse often governs the sourcing strategy. For example, a sales transaction's fact table can capture individual transactions on a given day. So the granularity becomes a transaction level granularity. The granularity of the fact table is also dependent on the granularity of source data. In this case the point of sales (PoS) would have the requisite granularity of data. In some situations the source data may be at a higher level of granularity while the downstream system such as the data warehouse might be requiring an aggregated value of the transactions on a given day.

Granularity of data from a reporting or analytics standpoint defines the data granularity needs of the data warehouse or data marts. However there are other considerations that also come into play such as -

- Data volumes and storage costs—choosing an appropriate level of data granularity affects the data volumes in the system of record or data warehouse. Higher volumes of data also involve greater data retention needs, which in turn imply higher storage costs. Hence the choice of data granularity in data warehouses and data marts becomes very important. Even though storage costs are not such a showstopper it is important to note that a small change in granularity can change data volumes significantly. Let us look at the following example:

  Average number of transactions per account in a retail bank = 50 per month

  Number of bank accounts in the given retail bank = 300,000

  Number of transactions in a given month = 300,000 * 50

  If the granularity changed to a month end snapshot whereby there would be one record per account, there would be a significant reduction of data volume and storage space.

- Query performance—Another important consideration in the granularity of the system of record or data marts is the query performance. Here it is important to consider the frequently executed queries and analyze the expected queries in the reports. Based on this analysis and the data storage needs the granularity can be defined. However it also is important to consider that in case of ad hoc reporting needs the ability to answer different types of queries would require data at a more detailed level. In case of less detailed data or less granular data the ability to perform ad hoc reporting diminishes to a great extent. In many cases queries can access tables with mixed granularity, for example, a detailed transaction fact table and a summary fact table.

- Source data availability—Although reporting needs and data storage needs may define data granularity of a data warehouse or SoR, one of the key drivers is the source data availability. There may be a need for a specific granularity of information in the SoR. However if the source systems do not have this granularity of information as needed by the SoR, then the granularity needs of the SoR will need to be compromised.

- Batch performance impact—Although it is advisable to have data in the system of record at the maximum level of detail or lower granularity; there are other considerations such as batch performance. Lower granularity means more events and transactions to capture, resulting in a higher volume of data that needs to be processed on a daily or intra-day basis. Higher data volumes mean a slower processing time window that can lead to business service level agreements not being fulfilled, which can result in reporting delays. As decision support systems look at trends and patterns, it is better to keep a granularity that is fairly detailed but does not mimic the source system granularity of information. Operational data stores can be designed to keep the same level of granularity as operational systems to aid in operational reporting and analytics, but this should not be the principle followed for data warehouses/SoRs.

■ ■ ■

CHAPTER 5

Pillar No. 2: Information Integration and Exchange

With an understanding of information sourcing, I now move to the topic of information integration and exchange. As shown in Chapter 4, information sourcing involves looking at the extraction strategies that are now extended to the integration approaches in vogue today. Integration approaches are largely driven by the target on which the data is loaded. For instance a data warehouse integration strategy will differ from that of an operational data store or a data integration hub. The nature of data, the latency of data, and the usage of integrated data play a key role in determining the data integration strategy. Information integration and exchange is a key tenet of EIM and deals with the information flow from source systems to consuming or subscribing systems using an integration engine. In today's complex business environments where a lot of data is exchanged between trading partners and suppliers, the integration strategy becomes crucial in addressing information gaps as well as a better understanding of business process efficiencies and bottlenecks. Some of the key drivers to consider while deciding on the integration approach include the following:

- *Nature of extraction between source systems and consuming systems (push/pull)*

- *Type of connecters needed for pull from source systems*

- *Leverage data integration engine for transformations of source data or use database engine for transformations*

- *Outbound extract formats needed for consuming applications*

- *Understand data security's needs as part of the integration process and any country's specific data compliance needs*

In the following sections I discuss each of these drivers in more detail.

Key Drivers for Determining Integration Approach

My goal for this chapter is to explain the key strategies and mechanisms concerning information integration and exchange, and to help you understand how information exchange standards are needed in business-to-business and business-to-consumer networks, and to explain the key tools for information integration and exchange.

Nature of Extraction Between Source Systems and Consuming Systems (Push/Pull)

If the agreed extraction mechanism from the source system is push based then the source system generates an extract on a predefined basis and transfers the file to a file landing area (secure area where the file can be landed for further processing). The data integration process polls the file landing folder where the source extract file arrives and then processes it for transformation and loading into the target system or consuming applications. In the case of a predefined time of source extract file, a pick and integration process would kick start at the defined time. When a pull is the agreed extraction mechanism, the integration process also has to write the extraction logic, execute the query on the source system tables, and then process the data. As is evident in the pull mechanism the integration process needs to be given access to source system tables to execute the query. Enterprises are often averse to providing direct access to operational systems such as ERP or CRM that run the business. In such a scenario the integration process would be provided access to replica source database (replicated in near real time) or in the case of ERP systems a set of interface tables.

Type of Connectors Needed for Pull from Source Systems

As discussed in the previous driver, the pull based mechanism would involve the integration process accessing the source system tables. In the case of database access such as SQL servers (e.g., DB2, Oracle) the open database connectivity (ODBC) connector would suffice. However for packaged applications such as ERP or CRM, application specific connectors are needed, for example, the connector for SAP or the connector for Siebel. The nature of source data also determines whether a connector is needed for not. For instance, if manual data is being integrated, no connector is needed as comma separated values (CSV) files or flat files can be integrated. This is a crucial part of the integration approach and awareness of the types of source connectors needed is a part of the data integration licenses procured by an enterprise.

Leverage Data Integration Engine for Transformations of Source Data or Use Database Engine for Transformations

There are two fundamental approaches to data integration namely 1) extract, transform, and load (ETL) or 2) extract, load, and transform (ELT). In the case of ETL, data is extracted from source systems, then transformed into load ready form using the transformation/business rules and then loaded into the target database. In ELT, the source data is extracted and then loaded into the target database where transformation to source data is completed using the database engine of the target database. The benefit of ELT is that transformation happens inside the database and the transformed data does not have to be sent across the network as is the case with the ETL approach.

Outbound Extract Formats Needed for Consuming Applications

The integration approach also needs to consider the outbound extract formats needed by the downstream/consuming systems. The integration job design needs to factor in the type of outbound extracts that will be generated for consumption by downstream applications and operational systems. In the case of data sent to external parties, such as suppliers or trading partners, the extract formats need to be defined and the interface agreements need to be in place.

Understand Data Security's Needs As Part of the Integration Process and Any Country's Specific Data Compliance Needs

Data security is a crucial part of the integration process and needs to be planned as part of the data integration design. The source systems need to send extract files using secure file transfer methods, and security needs to be designed for the file landing zone. Security ensures that a specific human resources source system transfers the extract files to a specific folder where the file transfer process has access. Likewise, the data integration process is given access to only the required folders in the file landing area. This ensures that data breaches do not happen as part of the integration process.

Information Integration and Exchange— Key Strategies and Mechanisms

Information integration and exchange is one of key solution components in the overall EIM landscape. The integration backbone ensures how effectively data is moved from the sources or the point of origin to the point of consumption and decision making. In this section I cover the different information integration strategies and their mechanisms. Although the focus is on data integration approaches, I also touch on enterprise application integration as a supplemental approach. The key integration strategies to be covered are as follows:

- Extract, transform, and load (ETL)
- Extract, load, and transform (ELT)
- Data integration hubs
- Slowly changing dimensions
- Real-time data integration
- Enterprise information integration

Extract, Transform and Load (ETL)

ETL is the standard data integration approach wherein the source data is extracted (either push or pull mechanism), transformed to a load ready form using business or transformation rules, and then loaded into the target database. The key points to note are: There is transfer of data over the network from source systems to the file landing area from where the integration process picks up the file and then loads into staging. Transformation rules are applied in the staging layer and then the data is moved again over the network and loaded into the target database (SoR, data mart, or operational data store).

Extract, Load and Transform (ELT)

ELT is a paradigm shift from the standard ETL approach. In this approach the source data is extracted (either push or pull mechanism), loaded directly into the target, and then the transformations are performed in the database. In the case of ELT there is only one data transfer from source to staging while in case ETL there were at least two data transfers from source to staging and staging to target. The comparison between the ETL and ELT approaches is shown in Table 5-1.

Table 5-1. ETL vs. ELT—Different Integration Approaches

ETL	ELT	Observations
Data transfers in the case of an ETL approach is between 1) source to staging and 2) staging to target.	Data transfer in the case of an ELT approach is one time between source and staging.	Data transfer over the network can potentially slow the data integration process. This makes ELT more efficient in terms of ETL load performance.
Transformations in ETL are completed in the data integration/ETL engine.	Transformations in the ELT approach are completed using the inherent capabilities of the database engine.	Transformations can be completed both in the database engine as well as in the data integration engine. However in the case of ETL the data needs to be transferred over the network making the process slower.
Time to market in the case of ETL transformations is quicker as the data integration engine comes with a number of built-in transforms.	Time to market in the case of ELT is slightly longer as the transformation largely uses database capabilities, which often implies that more queries need to be written. Even with ETL with its pushdown optimization, the transformations are performed in the database engine.	Time to market is slightly longer as testing needs to be done to test the data integration engine capability to load source data into database and then test the transformation logic in the database where transformations take place. In older ELTs, tool-stored procedures and queries used to be executed, which also needs to be tested.
ETL load performance	The ELT approach results in better data integration throughput resulting in faster data load times.	ELT enhances performance of data load by reducing the amount of data being moved from source to target and it leverages existing computing resources in the database.

There is no rule of thumb for determining, which is a better approach and in most cases a combination of both ETL and ELT approaches are adopted by enterprises. Hence it is recommended to go with a data integration tool that can support both types of architecture.

Data Integration Hubs

Organizations have traditionally approached data integration in an application or project specific manner. Project teams have focused on their specific requirements, built tightly bound, point-to-point interfaces between data sources and targets of interest. They built these interfaces without a good deal of regard for future adaptability and without considering the data integration processes developed by other project teams. This occurred due to the diversity of different use cases and project types needing data integration work (from data warehousing and business intelligence to interenterprise data sharing) and the distribution of teams involved in designing and building the data integration solutions.

As a result, organizations experienced higher deployment costs due to a lack of reuse; higher costs of ongoing support and enhancement due to extreme complexity; and higher levels of risk from a lack of visibility of data lineage and the potential for inconsistency in business rules for data transformation, calculations, and data quality controls. Data integration hubs are a delivery approach in which data integration processes are managed in a cross project/application fashion preferably by a data integration or data management center of excellence (COE; see Chapter 12 for more details on the CoE model). In a hub based model to data integration, data that is extracted from multiple sources flows through a centralized model (the hub) and is delivered from the hub to consuming applications (spokes). In effect, the hub serves as a clearinghouse for data moving between the combinations of sources and targets. Data may flow through the hub on a scheduled basis, batch basis, or a real-time and granular fashion (see Figure 5-1).

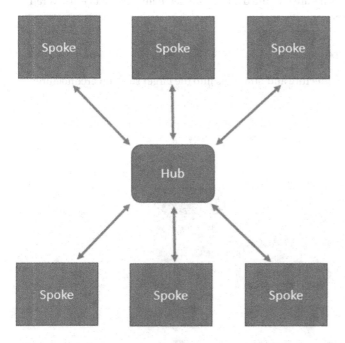

Figure 5-1. *Data integration hubs*

Key considerations for data integrations hubs are as follows:

- Persistence of data flowing through the hubs
- Canonical forms
- Data quality controls

Persistence of Data Flowing Through the Hubs

Enterprises may choose to make the hub a location for persistent data that flow through the enterprise. By persistence the hub becomes similar to an operational data store (ODS). In some cases the hub can be used only as a transient data store with storage that does not exceed seven days of history.

Canonical Forms

The data integration hub, irrespective of whether the data is persisted, can adopt forms and concepts of common formats that can be used as translation points between the differing syntax and semantics of the participating sources and targets.

Data Quality Controls

The centralization of data flow through the data integration hub creates the opportunity for a consistent approach to monitoring and improving the data quality. Data quality rules can be applied to data flowing through the hub, ensuring that data quality issues are identified and addressed before the data is delivered to the consuming applications and processes.

Data Integration Hub Architecture

Figure 5-2 shows the data integration hub architecture and how it fits into the larger solution architecture.

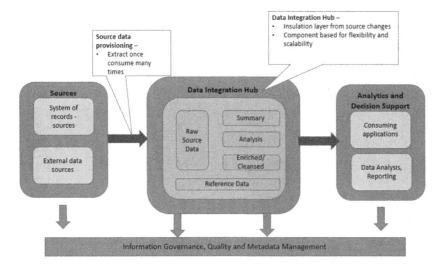

Figure 5-2. *Data integration hub architecture*

The source data can be from both systems of record sources (internal) as well as external sources. The raw source data is persisted in the hub for data exploration and raw source data needs for consuming applications; it also is cleansed/enriched, transformed, aggregated, and summarized. Reference data and master data also can be processed through the hub for consumption by master data management/reference data management hubs. This makes the hub the data ingestion engine for all enterprise analytics and data management needs. Gartner also recognized the hub as a key integration approach in its report 'Data Integration Hubs: Drivers, Benefits and Challenges of an Increasingly Popular Implementation Approach' and mentioned that there was growing interest among companies with regard to the integration hub concept. Some of the key benefits anticipated from adoption of data integration hubs include reduction in the number of data oriented interfaces, leveraging of canonical data forms through the hub construct, reuse of business rules for transformation, opportunities to enhance data quality, consolidation of infrastructure for data integration workloads, and enhanced understanding of data flow and lineage through end-to-end metadata.

Barriers to Adoption

Although there is no doubt about the potential benefits a data integration hub can provide, there are certain barriers to their adoption enterprises need to be aware of including

- Individual teams would need to give up their rights on decisions concerning the choice of tools, architecture, and design approaches.

- More resistance to change within an organization

- Funding issues based on defined business case and acceptability of charge back models (charge back models are commonly used in enterprises where a common technology infrastructure or service can be consumed by multiple departments or functions. The consumption drivers indicate how much charge back needs to be applied for a specific function or department.)

The best practice is to define a business case that avoids the questions concerning investments in data integration hubs especially where integration service lines/competencies already exist.

Slowly Changing Dimensions

Often in decision support systems and analytical data marts there is a need to track the historical changes in dimension attributes over time. For example the sales hierarchy or organization of an enterprise can change over time. When the hierarchy changes, there may be a requirement from business to track sales and report sales based on the both the historical hierarchy structure as well the latest hierarchy structure. This requirement can be addressed by implementing slowly changing dimensions (SCD) in the dimension tables. It is important to note that in the same dimension table there may be multiple types of SCD implemented. The major types of SCDs are as follows:

- Type 0—Type 0 is a passive method for tracking SCDs. No effort is made to track dimensions changes over time. So it is not possible to determine if the dimension attributes have overwritten values or have the historical values. This is rarely used.

- Type 1—In Type 1 SCD, no history is retained for the dimension attributes. Whenever the attributes values change, it overwrites the previous values. Type 1 SCD is used for attributes when there is no requirement to hold or retain history over time. From an implementation perspective, surrogate keys are used to optimize performance on joins as surrogate keys are identifiers defined by the database and are integers.

- Type 2—In Type 2 SCD, the complete history is retained by creating multiple records for the same natural key with different surrogate keys. Thereby unlimited history can be retained. Care must be taken with this style of implementation as the size of the dimension tables can grow explosively. SCD records can be tracked either by version number or effective date columns.

- Type 3—In Type 3 SCD, changes to the dimension attributes is tracked by adding columns and preserves only a limited history. Type 3 is limited in its ability to preserve history by the number of columns meant for storing historical data. Here is an example, Rohit is a postgraduate student and his home address has changed due to his parents moving from Mumbai to Delhi. So if we track *Address* as Type 3 SCD, there are two columns—Original Address and Current Address. If Rohit changes his address again then this design does not suffice as it does not retain the entire history.

- Type 4—In Type 4 SCD, there are two tables, one for holding current data and the other for storing history. Hence all changes over time are stored in the history table whereas the current table stores the most recent value of the dimension attribute. See Figure 5-3 for details. So if we look at Product dimension, there would be two tables—Product and Product history. Let's assume that the Product category has changed over time, so the Product table will have the most recent category whereas the Product history will have the historical classifications of the category with date time stamps for the period associated with the given category. This has some benefits

  - The entire history can be tracked for the dimension attributes tagged as SCD Type 4.

  - To avoid ETL bottlenecks the history table has all the previous attributes values.

  - Querying also is efficient as all reports seeking current values can be accessed through the main Product table. Only in cases where historical changes need to be reported as well will they be accessed along with the main dimension table.

Product table

Product Key	Product Code	Product Name	Product Category
123	CLA12	Cola	Beverages
124	BK11	Baking Soda	Baking Products

Product History

Product Key	Product Code	Product Name	Product Category	Create Date
123	CLA12	Cola	Aerated Drinks	26-Jun-2011
123	CLA2	Cola	Carbonated Drinks	14-Apr-2013
124	BK11	Baking Soda	Bakery & Food	14-Aug-2013

Figure 5-3. *Type 4 slowly changing dimensions (SCD) illustration*

- Type 6 or hybrid—Type 6 or hybrid SCD combines approaches of Types 1, 2, and 3 (1 + 2 + 3 = 6). Here is an example to explain how Type 6 works and is implemented. The customer dimension structure is given in Figure 5-4.

Customer table

Customer Key	Customer Code	Customer Name	Current City	Historical City	Start Date	End Date	Current City Flag
100	G123	Ravi Gupta	Pune	Pune	02-Feb-2009	31-Dec-9999	Y

Figure 5-4. *Customer table*

A customer moves from one city to another (say from Pune to Bangalore). A new record is inserted for the customer (Type 2) in the customer master/dimension table. The new Current City and Historical City in the new record are set as Bangalore and Pune, respectively, with the Current City of the previous record updated to "Bangalore" (Type 1). There is an additional column called Historical City to capture the previous city of residence (Type 3). See Figure 5-5 for details.

Customer table with city changes

Customer Key	Customer Code	Customer Name	Current City	Historical City	Start Date	End Date	Current City Flag
100	G123	Ravi Gupta	Bangalore	Pune	02-Feb-2009	09-Aug-2012	N
101	G123	Ravi Gupta	Bangalore	Bangalore	10-Aug-2012	31-Dec-9999	Y

New record inserted for Type 2

Type 1 – Current city updated in previous record as well

Extra column to maintain previous value of city (Type 3)

Figure 5-5. Type 6 slowly changing dimensions (SCD) illustration with different surrogate key

One disadvantage in this particular implementation style is that each time the customer attribute changes, a new surrogate key is inserted into the customer master whereby the master data changes, which has an impact on the transaction data as well. There is another implementation approach in which the surrogate key does not change and the table looks like Figure 5-6.

Customer table with city changes

Customer Key	Customer Code	Customer Name	Current City	Start Date	End Date
100	G123	Ravi Gupta	Pune	02-Feb-2009	09-Aug-2012
100	G123	Ravi Gupta	Bangalore	10-Aug-2012	31-Dec-9999

Figure 5-6. Type 6 slowly changing dimensions (SCD) illustration with same surrogate key

This is clearly a cleaner way to implement the changes in the customer dimension and makes it easier to query the associated fact tables.

Real-Time Data Integration

Often in enterprises there is a business need to integrate business information in real time or near real time to provide operational insights or analytical requirements such as analyzing transaction patterns to govern fraudulent activities. The traditional data integration tools have evolved over time to provide capabilities to perform real-time or near real-time data integration capabilities. The principal approaches to real-time data integration include the following:

- Change data capture

- Events or streams based integration

Change Data Capture

CDC is an integration approach based on identification of changes to source data, the capture and delivery of these changes to the consuming systems. The benefit of the CDC approach is that only the changed data (inserted/updated/deleted) is moved, thereby making the process more efficient and faster. The key benefits of this approach include a) enhanced business agility through faster data integration, b) improved IT response to near real-time business data needs, and c) reduced IT costs through reduced usage of resources. However CDC works primarily with structured data sets. There are multiple approaches to CDC and the common ones are, 1) time stamps on rows where the last changed date gives the time of the last change, where delta changes can be picked up. 2) Version number in rows ensures that data related to the most recent version number is the only version picked up. 3) Triggers based approach involves triggers (delete/update/insert) that capture changes in the transaction tables and log these changes into queue tables. The queue tables are then consumed by target systems. 4) Log scrapping are changes made to database contents and metadata that are recorded in the database's transaction log. With log scrapping it is possible to capture the changes made to the database. However with CDC based on log scrapping the technique becomes database specific as transaction logs are database specific. Another challenge is that log scrapping means that CDC agents need to be installed in the production transaction system resulting in some performance degradation in the transaction systems. A viable alternative is to replicate the production transaction system to another backup/replica server and apply the CDC on the replica server, thereby not impacting the performance of the production systems.

The key benefits of going with a log scrapping CDC are as follows:

- There is no need to program the capture of the changes in the transaction system.

- There is low latency in getting the changes, thereby enabling faster integration and quicker decision making.

- Transaction integrity is not compromised as the log scrapping produces the same stream in which the transactions were executed.

- There is minimal impact on source applications.

I now analyze some of the common integration patterns that come with CDC solutions. I look at four integration options.

- CDC with staging database—As transactions happen in the source application, the CDC engine captures these changes and writes these changes on to a staging database. The ETL tool reads these changes from the staging table and loads these changes to the target database (consuming application—ODS, SoR, or business application). See Figure 5-7.

Figure 5-7. *CDC with staging database*

- CDC with middleware—As transactions happen in the source application, the CDC engine captures the changes and writes these changes to a middleware engine (queue tables). The ETL tool connects to the middleware engine using a connector and then loads these changes into the target database (consuming application—ODS, SoR, or business application). Refer to Figure 5-8.

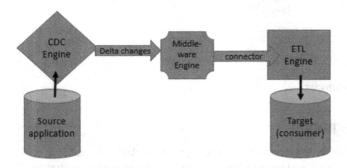

Figure 5-8. *CDC with middleware*

- CDC with flat files—As transactions happen in the source application, the CDC engine captures the changes and writes these to a flat file. The ETL engine reads the changes from the flat file and loads these changes into the target database (consuming application— ODS, SoR, or business application). Refer to Figure 5-9.

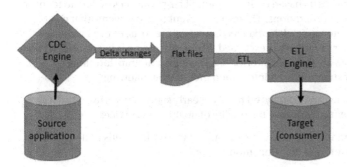

Figure 5-9. *CDC with flat files*

- CDC with ETL engine direct connector—Transactions occurring in the source application are captured by the CDC engine. The changes are passed on from the CDC engine to the ETL engine through a connector from the ETL to the CDC engine. The ETL engine processes the changes and loads them into the target database (consuming application—ODS, SoR, or business application). See Figure 5-10.

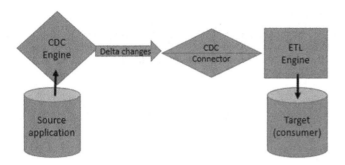

Figure 5-10. *CDC with ETL engine direct connector*

Events and Streams Based Integration

Although there are numerous data integration technologies to enhance the speed of data processing, a few disruptive integration solutions such as events and streams based integration are now gaining ground. Given the rise of big data processing needs and complex business requirements such as responding to events in real time, managing heterogeneous data, and adaptability to changing data forms has resulted in a new approach to data integration. Events or streams based integration allows queries to be made in real time, which is a paradigm shift from traditional integration methods where data is loaded and then queries are executed on the history and current data. In events processing data are combined from various sources and inference is drawn about patterns or events. The objective of event processing is to detect events such as opportunities and threats and respond to them as quickly as possible. Enterprises can respond to opportunities and threats with near real-time alerts based on integration of large volumes of data to identify business events of interest based on predefined conditions. There are different types of events that can be detected including sales leads, orders, news items, stock market feeds, weather reports, or even a change of state where threshold values of a particular parameter is exceeded. Event processing is based on a set of technologies such as event pattern detection, abstraction, filtering, aggregation, and transformation.

Some of the key use cases for events or streams based integration include the following:

- Health care—Medical device data can be analyzed to detect early signs of disease, correlations of data across patients to improve the quality of health care services.

- Transportation—Helps city planners and operators to understand traffic trends and efficiency of transport services through real-time monitoring.

- Telecommunications—Analyze call detail records in near real time to analyze customer behavior and propensity to move to other providers as well as response to promotions.

Enterprise Information Integration

Enterprise information integration (EII) is an integration mechanism by which heterogeneous data sources appear to the business users as a single homogenous source. Some of these common approaches include data federation and data virtualization. EII systems have two primary patterns of integration:

- Mediation—This is where the EII system acts as a broker between multiple applications (e.g., ERP and CRM or business intelligence systems). When an event occurs (of business interest) the integration component of the EII system is notified and the changes are propagated to the consuming applications.

- Federation—Here the EII systems acts as facade over a set of business applications, such as CRM, SFA, or ERP, and any request for information from other applications to them is handled by EII. The EII system provides only the requested information by providing interfaces to the requesting applications.

Information Exchange Standards—Business to Business and Business to Consumer

After looking at the different data integration approaches I now analyze how enterprises exchange data with other businesses and customers. There are numerous information exchange standards that are largely based on business-to-business (B2B) information exchange standards and there are a few additional standards around business-to-consumer (B2C) information exchanges. The need for standards is to ensure that information exchanged is governed by certain criteria and protocols that are commonly understood and involve a common business glossary and metadata. I first visit the B2B standards for information exchange. Some of the common B2B standards are shown in Table 5-2.

Table 5-2. *Business-to-Business (B2B) Information Exchange Standards*

Industry	Information Exchange Standard	Comments
Insurance (life insurance, property and casualty, reinsurance and commercial)	ACORD—the Association for Cooperative Operations Research and Development—they define the electronic data standards for the exchange of insurance data between trading partners.	The common standards include ACORD XML for life insurance, property and casualty insurance and reinsurance and large commercial. Information exchange is based on predefined XML formats.
Petroleum	PIDX—Petroleum Industry Data Exchange is the American Petroleum Institute's committee on global electronic business standards in the oil and gas industry.	PIDX provides several XML specifications that support various aspects of the oil and gas supply chain. EDIX12 and FTP standards also are defined.
Pharmaceutical	CDISC—Clinical Data Interchange Standards Consortium develops standards to support electronic acquisition, exchange, submission, and the archival of clinical trials data and metadata for medical and biopharmaceutical product development.	CDISC operational data model— XML specification format for interchange and the archival of data collected in clinical trials. The model represents study metadata, data, and administrative data associated with a clinical trial.
Health care	HL7—Health Level Seven develops standards for the exchange, management, and integration of data that supports clinical patient care and management, delivery and evaluation of health care services.	HL7 XML format—XML specification for the exchange of clinical data and information. The purpose of the exchange of clinical data includes, but is not limited to, provision of clinical care, support of clinical and administrative research, execution of automated transaction oriented decision logic, support of outcomes research, and so forth.

(continued)

Table 5-2. (*continued*)

Industry	Information Exchange Standard	Comments
Engineering and Manufacturing	ISO 15926 is an 11 part model and data library reference for the process plants and production facilities. It is the standard design for data integration, sharing, and exchange between computer systems for the information of the life cycle on the process plants, including oil and gas production facilities.	Ten parts of ISO 15926 standard are: Part 1 (engineering, construction, and operation of production facilities), Part 2 (data model), Part 3 (reference data for geometry and topology), Part 4 (reference data for the process industry), Part 7 (integration of life cycle data), Part 9 (implementation standards), Part 10 (test methods), Part 11 (industrial usage guidelines), Part 12 (life cycle ontology), Part 13 (integrated life cycle asset planning).
Retail	ARTS—The Association for Retail Standards define the focus on A2A (application-to-application standards). ARTS has four standards.	Retail data model, unified point of service (unifiedPOS), IXRetail XML schemas to integrate applications within the retail enterprise and standard requests for proposal (RFPs) to guide retailer selection of applications and provide a development guide for vendors.

Some of the common B2C standards are shown in Table 5-3.

Table 5-3. *Business-to-Consumer (B2C) Information Exchange Standards*

Industry	Industry Exchange Standard	Comments
Finance	IFX—Interactive Financial Exchange	XML specification for consumer-to-business payments and consumer-to-business banking.

IFX also is used for B2B data exchange as well. To illustrate how common information exchange is between B2B mainly and B2C there are some industry standards provided. There are numerous other information exchange standards that can be found in industry specific books and articles.

Tools for Information Integration and Exchange

After reviewing the information integration approaches and the different information exchange standards by industry, I now move to the tools prevalent in the market today. The data integration market has now evolved and there are data integration suites that provide a vast multitude of functions from CDC and data replication; data profiling; data cleansing; extract, transform, and load; or extract, load, and transform; streams or events based processing.

Some of the market leading tools in each of these categories are covered in Table 5-4.

Table 5-4. *Common Data Integration Tools (Based on the Gartner Magic Quadrant 2014)*

Data Integration	Informatica	IBM	SAP
Extract, transform, and load/extract, load, and transform	Informatica PowerCenter	Infosphere DataStage	SAP BusinessObjects Data Integrator
Data profiling	Informatica Data Explorer	Infosphere Information Analyzer	SAP BusinessObjects Data Services
Data cleansing	Informatica Data Quality	Infosphere QualityStage	SAP BusinessObjects Data Services
Change data capture	Informatica PowerExchange Change Data Capture Option	Infosphere Change Data Capture	SAP Sybase Replication Server and SAP BusinessObjects Data Services
Events/streams processing	Informatica Vibe Data Stream	Infosphere Streams	SAP Event Stream Processor

■ ■ ■

CHAPTER 6

Pillar No. 3: Information Governance and Quality

With an understanding of information integration and sourcing, I now move to one of the horizontal capabilities in enterprise information management. Horizontal implies a capability that cuts across the entire lifecycle of information from data sourcing to the delivery and consumption of information for reporting and analytics. Information governance is one of the key pillars of enterprise information management and it is not about technology. It involves people taking the responsibility for information assets across the organization by reviewing the processes they leverage to interact with information as well as how and why it is used.

Creating an information governance framework to ensure confidentiality, quality, and integrity of data is essential to meet both internal and external requirements such as regulatory compliance, financial reporting, data security, and privacy policies. Effective information governance eliminates risk, both business and compliance risk, by introducing controls that introduce oversight into the information creation, exchange, and transformation processes. Information governance enables enterprises to integrate and consolidate information from different lines of business and functions into a single "trusted" version of truth, providing economies of scale and making it possible to effectively tie information policy to business strategy. Although on paper it appears a no brainer that the need for information governance is greater than ever before, the reality is that many information governance initiatives have been bogged down in bureaucracy and have not delivered the expected outcomes. In many cases senior management is not completely "on board" due to ineffective or poorly defined business cases, which results in information governance being an IT funded initiative. It is imperative for strong senior management support and their buy in for information governance initiatives to succeed at enterprises. The following sections cover these key considerations for information governance.

- *Define information governance and quality*

- *The key drivers for information governance and quality*

- *Building blocks for information governance and quality*

- *Critical success factors for information governance and quality*

- *Tools for information governance and quality*

Now I discuss each of these key considerations in more detail.

■ **Note**   This chapter's goal is to explain what is information governance and quality, the key drivers for information governance and quality, and how organizations can go about building the foundations for effective information governance and quality.

Although I have introduced the concept of information governance, it is also important to consider the quality of the information based on what enterprises classify as data, and take decisions that impact the business cycles. So although information governance brings in the required oversight and controls, information quality brings in the quality and cleansing processes and tools that ensure that information sourced, integrated, and transformed within the enterprise for use in business processes and decision making is of an acceptable standards. There are numerous information quality standards by industry that I briefly touch on in the next section.

Define Information Governance and Quality

Although there are numerous definitions for *information governance*, the most practical one would be—to manage, enhance, and maintain the quality of information that an enterprise generates and consumes. Information governance unites people, process, and technology to transform the way information assets are sourced, managed, maintained, enhanced, and shared across an enterprise as well as with trading partners and leveraged to enhance profitability and business strategy execution.

Gartner defines information governance as the specification of decision rights and an accountability framework to encourage desirable behavior in the valuation, creation, storage, use, archival, and deletion of information. It includes processes, roles, standards, and metrics that ensure the effective and efficient use of information in enabling an organization to achieve its goals.

The Data Warehousing Institute (TDWI) defines information governance as an executive level information governance board, committee, or other organizational structure that creates and enforces policies and procedures for business use and technical management of information across the entire organization. Common goals of information governance are to improve information quality, remediate its inconsistencies, share it broadly, manage change relative to data usage, and comply with internal and external regulations and standards for information usage.

Information quality is often the quality of information in the business applications of the enterprise. From a data warehousing and business intelligence standpoint, *information quality* refers to the quality and "fit for use" capabilities of information stored in the data warehouse and other decision support systems. Information quality has numerous frameworks and the widely accepted common dimensions of information quality are—accuracy, consistency, timeliness, completeness, reliability, security, and accessibility.

The building blocks of information governance and quality include the following:

- Information governance processes
- Information governance council
- Information governance tools
- Information quality processes
- Information quality organization model
- Information quality tools

Information Governance Processes

Information governance processes are defined as part of setting up the framework to support the governance of information assets in any small or large enterprise. Information governance processes help to classify the information in terms of sensitivity, compliance, enterprise risk, and financial impact. They also help to derive the information consumption needs of business functions and units and provide a data access framework that governs who is allowed to see which information. Information governance processes also put in place, audit and control mechanisms to monitor whether data breaches are happening and information shared with external parties are in line with the data security policies of the enterprise and industry in question.

Information Governance Council

Information governance council is the executive body that defines and approves the information policies and endorses the audit policies concerning the information governance in an organization. The information governance council is staffed by senior business leaders (COO, CXO) as well as chief information officer (CIO), chief information governance officer (CIGO), among others. In some organizations it is also known as the data governance board.

Information Governance Tools

Information governance tools include data discovery tools, business glossary, and metadata tools that help disseminate the common terms used enterprise wide, as well as master and reference data governance tools, data archival tools, data masking tools. These tools are covered in a later section.

Information Quality Processes

Information quality processes are defined as part of setting up the quality framework that ensures that the quality of information sourced, integrated, transformed, and consumed within an enterprise is trustworthy, recent, consistent, and integrated.

Information Quality Organization Model

The information quality organization model is described to ensure that the information quality processes are defined and followed within the enterprise in line with the information governance process framework in place. The information quality organization model includes strategic, tactical, and operational levels with people identified to staff roles at each of these levels.

Information Quality Tools

Information quality tools include data profiling tools, data cleansing and enrichment tools that are used to profile as well as apply data cleansing rules to the source data being extracted to ensure that data being sourced and integrated is of an acceptable standard. These tools are covered in a later section.

Some of the commonly used industry standards for information quality include:

- *ISO 8000 the international standard for data quality*—This is the ISO standard for data quality and is comprised of the following parts: Part 1 (overview), Part 2 (vocabulary), Part 100 (master data—exchange of characteristic data overview), Part 102 (master data—exchange of characteristic data: vocabulary), Part 110 (master data—exchange of characteristic data: syntax, semantic encoding, and conformance to data specifications and others). Although the 100 series of parts deal with master data other series of parts deal with transactional data, referenced data, and engineering data. Parts 1 through 99 are reserved for general data quality issues that include data governance, although this also may be covered in more detail in the other series. Additional reading - `http://iaidq.org/publications/doc/west-2009-07.shtml`.

- *Eurostat data quality definition*—Eurostat's mission is to provide the European Union with a high quality statistical information service. The Eurostat quality assurance framework (QAF) is embedded in total quality management and describes the tools and procedures put in place to ensure that the statistics produced are of high quality. The quality of statistical outputs is assessed against six criteria namely: relevance, accuracy, timeliness and punctuality, accessibility and clarity, comparability, and coherence. Additional reading - `http://ec.europa.eu/eurostat/web/quality`.

Key Drivers for Information Governance and Quality

Although information governance should be an enterprise wide initiative, it is often driven by business drivers such as master information management, information quality, operational, and compliance related risks. Enterprises often define a business case for information governance that in turn could be driven by an initiative such as cost optimization and lower transaction costs. Whatever the scope of the information governance initiative (enterprise wide, business unit, or initiative driven), the common business drivers for information governance and quality are shown in Table 6-1.

Table 6-1. *Key Drivers for Information Governance and Quality*

Key Business Driver	Rationale
Master information management	One of the common drivers for information governance is master information management programs. Enterprises often grapple with master data scattered over a plethora of systems leading to higher transaction costs and higher order execution times as well as invoicing errors.
Mergers and acquisitions	Mergers and acquisitions often result in two enterprises with different organization hierarchies and master data entities. This results in the revamping and restructuring of master data entities. This often is driven by an information governance program to ensure consistency of enterprise definitions, business glossary, and metadata.
Single view of customer	Often a single view of customer initiatives drives master information management programs and this in turn drives information governance and quality initiatives.

(*continued*)

Table 6-1. (*continued*)

Key Business Driver	Rationale
Cost optimization	Cost optimization is often an enterprise or a business unit specific initiative that is driven from the top and requires high quality master and transaction data, which in turn are products of robust information governance, quality, and metadata management.
Operational risk	Operational risk arises due to multiple definitions of key master data entities as well as inaccuracies in reference data. This impairs an organization's ability to interpret business outcomes as well as results in operational risks due to data inconsistencies and reporting issues.
Compliance risk	In the absence of governance and quality, an enterprise often runs the risk of inconsistent reporting of mandatory/compliance reports resulting in fines and/or loss of investor confidence.

Building Blocks for Information Governance and Quality

Despite numerous efforts and investments in information management technology, most companies still struggle with information governance programs. The practice of managing structured data in corporate systems and data warehouses may have gained some maturity but with the surge in unstructured data sources in the enterprise, the discipline of information governance faces new challenges. It is crucial for enterprise to approach the problem in a structured fashion. The key enablers that help build information governance and quality in a large enterprise are: information governance policy and framework and information governance organization. Although policy and framework build the set of procedures to be followed for information assets generated within and outside the organization, framework helps build the information governance charter to define the key tenets or principles around information governance. Information governance organization is the organization structure that works at different levels to build governance in an enterprise (see Figure 6-1 for a sample information governance organization). The information governance organization is comprised of the following levels—executive level (includes CIGO or executive data sponsor), strategic level (information governance lead, BI Centre of Excellence (CoE) leader), tactical level (lead data steward, data quality lead, process owner), and operational level (source data steward, data quality specialists, and data integration specialists).

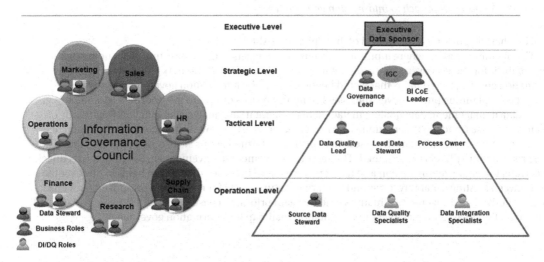

Figure 6-1. *Sample information governance organization*

The starting point for an enterprise is to define a problem statement and build a business case around the problem statement to fund the information governance and quality initiative. There is considerable debate whether these initiatives should be business unit focused or enterprise wide, the short answer to which is to start small and then build incrementally as business participation grows as the benefits of a robust governance framework and methods are realized. The key point is that information governance initiatives need to be driven by business, and there must be top management buy-in to ensure continued interest and success of these programs.

An information governance program should kick off with an assessment of the current state of maturity in an enterprise, to ascertain the gaps and build a road map with which to base the information governance journey. See Figure 6-2 for the phased approach to information governance.

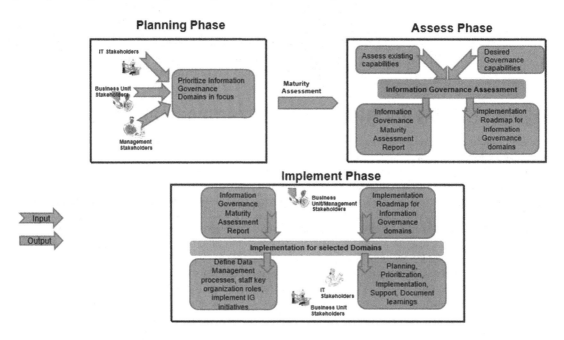

Figure 6-2. *A phased approach to information governance*

The phased approach is comprised of three phases as shown in Figure 6-2.

The planning phase is where inputs are taken from management, business unit (depending on the scope of the information governance program in question), and IT stakeholders to ascertain the key information governance domains that would be the focus of the information governance assessment.

Once the planning phase is over, the assessment phase kicks off where the prioritized domains (e.g., master data management, data quality, metadata management) are analyzed in detail through study of systems, processes, and documentation and through detailed interviews to analyze the existing practices and ascertain the gaps. The maturity assessment also is performed at this stage based on which the enterprise's maturity level is ascertained. The senior management involvement ensures that the feedback concerning the organization's vision and target maturity level is known that in turn helps to discover the gaps between existing capability levels and the target maturity level for each of the prioritized domains. The gaps derived are broken into initiatives that are again prioritized based on business value and ease of implementation to ascertain a target implementation roadmap for information governance.

In the implement phase, the key initiatives are implemented based on prioritization of initiatives based on business value and ease of implementation. It also involves staffing the key roles in the information governance organization at various levels depending on the skills needed to implement the initiatives. The projects also are tracked to ensure that the benefits expected are being delivered to ensure a feedback loop back into the business cases. The learnings are documented and the staff trained on data management processes to ensure greater acceptance and adherence to the best practices.

Table 6-2 discusses the activities related to each of the phases along with activities and deliverables.

Table 6-2. *Activities and Deliverables in a Phase Information Governance Engagement*

Phase	Activity	Deliverable
Planning	Identify the information governance domains in scope for the assessment, derive stakeholder map, prepare interviews and workshop schedules Perform maturity assessment	Information governance assessment project plan Interview schedule
Assessment	Assess current maturity level and capability gaps, define target state, define information governance road map and business case	Information governance maturity assessment report, information governance road map and business case
Implementation	Define information governance framework, council, information management policies, define information governance KPIs and reports, implement road map initiatives	Information governance framework, information management policies, implementation documents, information governance KPI reports and dashboards

With an understanding of the enablers, I now focus on the building blocks or core domains of information governance that include: master information management, information quality, information lifecycle management, information security, and metadata management. Master information management is covered in detail in Chapter 7; however, it is a key domain to consider when analyzing the information governance maturity of an enterprise. Often enterprises spearhead their information governance programs with key domains like master information management and information quality. Information quality programs can be driven by separate initiatives such as single view of customer, master information management, or by information governance programs as well.

When driven as separate initiatives, information quality programs also commence with a strategy and assessment phase to ascertain the key enterprise information quality issues that could be due to broken business processes, lack of governance in the processes, data quality issues in the source systems, as well as technology related issues. Information quality programs also include building monitoring processes to ensure that quality of information that flows between business applications does not degrade as it moves away from the point of creation. The overall information quality process is summarized in the steps shown in Figure 6-3.

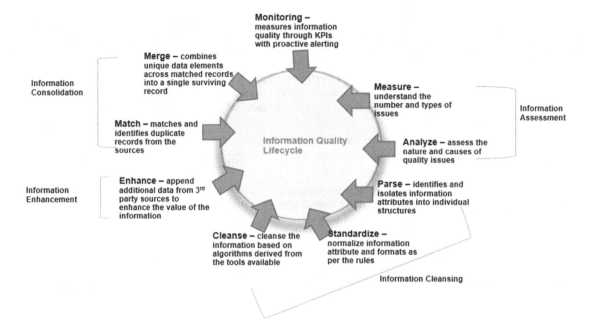

Figure 6-3. *Stages of the information quality lifecycle*

The information quality lifecycle can be classified into the following phases as shown in Figure 6-3. These include the following:

- *Information assessment*—This includes measuring the number and type of quality issues and analyzing the causes of the quality issues. The assessment phase involves the use of automated data profiling tools such as Information Analyzer from Informatica or Infosphere Information Analyzer from IBM.

- *Information cleansing*—This includes parsing the information into individual structures, standardizing the data as per the formats expected, and then cleansing the information by running it through the inbuilt algorithms provided by the selected cleansing tool.

- *Information enhancement*—Once the information is cleansed it can be further enhanced to increase its business value. For example, customer master information can be enhanced with demographic data purchased from external third party data providers to increase the value of customer information that can be used for targeted marketing campaigns.

- *Information consolidation*—After enhancement is finished the information is consolidated and the duplicates are identified and the surviving record can be determined. Some attributes that are more recent and reliable are merged from duplicate records into the surviving record by a merge process.

- *Continuous monitoring*—This is achieved by defining information quality key performance indicators (KPIs) across the information processing lifecycle to ensure quality is monitored continuously and issues can be reported proactively. Examples of KPIs include percentage duplicates in source files, file size ranges, number of KPIs with incorrect values, and so forth.

Information security is another key domain in information governance. This is more important with compliance needs in many industries and with global corporations dealing with data security breaches that impact reputation as well as consumer confidence. Information security is often driven as a separate initiative and also can be considered as part of a wider information governance program. Information security programs need to look at classifying information assets in an enterprise and then access the required controls to be placed over the information assets. The process of determining the correct mix of security and information access is determined based on an information consumption security process model. The information consumption security process model is comprised of two phases—discovery and mitigation. See Figure 6-4 for the discovery phase activities. The discovery phase includes the following tasks:

- *Classify information assets*—In this activity the information assets are classified based on the enterprise's definition of data confidentiality and business domain considerations. Information can be classified as highly confidential, confidential, internal, and public. Business domains can be classified as customer, employee, vendor, operational, and financial. For example, customer transaction value is financial data that is confidential data whereas the customer name and social security number are considered as highly confidential data and part of the customer business domain.

- *Analyze business function*—After completing the inventory of information assets, the next step is to derive the risk profile of each business function related to the information assets. The key is to identify the data elements accessed by a business function.

- *Define employee risk profile*—Based on the role of an employee and the business functions performed, a risk profile should be defined for each of the information consumers.

- *Analyze system function*—Performing any business function involves access to information systems. Hence to manage information security in an effective manner, a risk profile of each of the systems needs to be identified based on the information elements in the system.

- *Business to system map*—After completing the business function and system function, it is important to derive the combined risk based on business to system access level. The combined risk is derived by multiplying the business function risk with system risk.

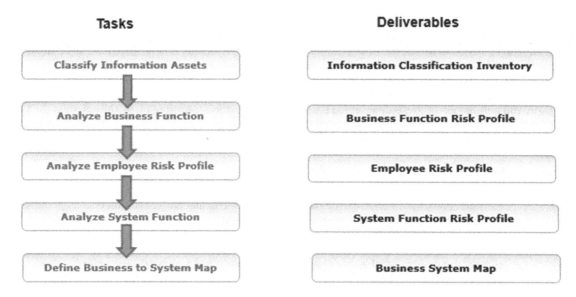

Figure 6-4. *Information consumption security process model—Discovery phase*

See Figure 6-4 for the associated deliverables for all the tasks mentioned in the discovery phase.

The associated deliverables for the discovery phase are the information classification inventory, business function risk profile, employee risk profile, system function risk profile, and the business system map.

The mitigation phase of the information consumption security process model (see Figure 6-5) is comprised of the following tasks:

- *Review access control*—Although most enterprises would have an access control policy in place, it is of paramount importance to analyze and assess whether the users have access to the right amount of information. Two very common scenarios to look out for are systems in place where very granular level security cannot be achieved and the use of generic IDs in groups where customized security needs for every user cannot be addressed. It is mandatory to understand the job function of each of the users and the data to which they have access. Based on this activity, organizations need to derive an access control inventory that will provide them with an insight into the users whose access control needs to be modified to map suitably to their job functions.

- *Audit systems*—Systems have to be audited on a frequent basis (once a month to once every three to four months) depending on the criticality of information stored in them. It has been observed that systems have additional information that has not been used by end users. Audits need to be performed on used as well as unused information to assess the security threat provided by unauthorized access to this information. Audits also look at regulatory needs (compliance test), access control (access test), and persistence of information (persistence test). The different types of test are described in the following:

  - Compliance test—Tests to verify whether the system in question complies with specific regulatory requirements such as the Sarbanes Oxley Act, HIPAA, and so forth, and corporate security policies.

  - Access test—Tests to verify whether a system requires access to specific information, such as does a marketing system need access to sensitive customer information, for example, a customer's social security number?

- Persistence test— A system often needs access to confidential data. A persistence test will test whether there is a need to store the confidential data in the system. The system in question needs access only during a transaction and does need to store such the value. Audits can help in determining which systems are violating the persistence test. When critical data needs to be stored in the system, it may need to be encrypted to ensure that even in case of a security breach the critical data is not exposed in a readable form.

- *Develop training plan*—One of the key aspects of information protection and security is to train employees regarding the importance of information security and how to protect information from any intentional or unintentional breach. Training plans need to be tailored based on the following factors:

  - Employee risk profile—Employees with a higher risk profile would need to attend more training sessions based on perceived risks.

  - Employee job function—Training needs to be customized depending on the employee job function.

  In addition training needs to be an ongoing process as the employee job function changes with time and although access needs to be changed, so does the employee risk profile and the corresponding training needs.

- *Develop a contingency plan*—Despite all efforts made to check information for security breaches, it is mandatory to have a contingency plan in place. A contingency plan should be comprised of the following aspects:

  - Communication plan for the employees in case of a breach

  - Communication plan for the affected persons or parties

  - Communication plan for the outside world (shareholders, stakeholders, public)

  - Damage control plan with law enforcement agency

  - System damage control plan

  - Impact analysis through system audit

Figure 6-5. *Information consumption security process model—Mitigation phase*

To ensure information security, enterprises need to analyze where sensitive customers', financials', and employees' information reside, who has access to this information, and how it has to be protected from potential breaches of security.

Information lifecycle management (ILM) is another key aspect of information governance in enterprises. Methods, practices, processes, and tools used by enterprises to manage information consistently and effectively over the course of the information lifecycle are covered under the gamut of ILM. Collection, creation, transformation, migration, distribution, utilization, access, and archiving, up to and including retirement/disposal of data according to regulatory requirements and enterprise needs are included. ILM involves defining policies around information storage and retention as well as archival and the removal of information assets from enterprise information stores. ILM involves defining standards for each phase of the information lifecycle and also processes the monitoring of information relevance and enforcement of policies. ILM strategy is a subset of information governance strategy and also involves choosing the right technologies to ensure information if governed and flows across the information lifecycle in compliance to regulatory needs as well as enterprise policies. Some of the common ILM tools used are for the purposes of data masking, data retention, and data archival. ILM policies and tools ensure that information that is accessed more frequent is stored on faster storage media (which costs more) whereas less critical or less accessed data is stored on cheaper and slower media. With increasing compliance standards that mandate critical information, such as customer details, health care data needs to be masked and encrypted when they pass over from one environment to another. Hence more emphasis now resides on an enterprise to define enterprise-wide ILM strategy and policies to support such needs.

Metadata management is another key pillar of information governance. Although it gets less recognition in the world of big data in which semantics of the data become extremely crucial to its understanding, metadata is once again considered as a true source of value. Metadata management is often one of the key drivers of an information governance initiative. Metadata is primarily one of three types—business metadata (consistent definitions of business terms, business glossary), technical metadata (collection of data model, ETL design artifacts, and BI measures and report definitions), and operational metadata (about ETL and BI processes). Metadata management is a key pillar for governance of information assets in an enterprise as it provides the following benefits (see also Figure 6-6).

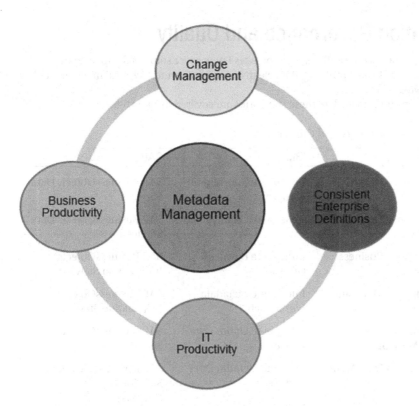

Figure 6-6. *The business and technical benefits of metadata management*

- *Business productivity*—Through data lineage and impact analysis capabilities provided by end-to-end metadata, businesses can know the accuracy, completeness, and currency of the data in the information data stores and business decision-making models.

- *Change management*—Enhances the management of data integration changes and reduces the errors caused by change. By providing enterprise-wide visibility into data definitions, lineage, and relationships, metadata management builds confidence in the data used to make critical business decisions.

- *Consistent enterprise definitions*—Business metadata helps to establish an enterprise-wide, business-centric vocabulary of mutually agreed, centrally governed, and shared business terminology.

- *IT productivity*—The availability of end-to-end metadata provides IT with the ability to predict the impact of data model changes or new business requirements, thereby enhancing IT productivity.

Tools for Information Governance and Quality

After reviewing the information governance and quality processes I now cover some of the tools in this domain. Some of the key tools include data profiling, data cleansing, business and technical metadata tools, data masking, and data archiving tools.

Some of the market's leading tools in each of these categories are covered in Table 6-3.

Table 6-3. *Tools for Information Governance and Quality*

Information Quality	Informatica	IBM	SAP
Data profiling	Informatica Data Explorer	Infosphere Information Analyzer	SAP BusinessObjects Data Services
Data cleansing	Informatica Data Quality	Infosphere QualityStage	SAP BusinessObjects Data Services
Business metadata	Informatica Business Glossary	Infosphere Business Glossary	SAP BusinessObjects Information Steward
Technical metadata	Informatica Metadata Manager	Infosphere Metadata Workbench	SAP BusinessObjects Information Steward
Data masking	Informatica Dynamic Data Masking	Infosphere Optim Data Privacy	SAP BusinessObjects Data Services
Data archiving	Informatica Data Archive	Infosphere Optim Data Archive	SAP BW NLS (Near Line Solution) Database
Data activity, monitoring, and audit	No direct tool	Infosphere Guardium Data Activity Monitor	Monitoring by AppDynamics (Sybase IQ)

■ ■ ■

CHAPTER 7

Pillar No. 4: Master Information Management

With an understanding of information governance and quality and the key role it plays in managing information assets in an enterprise, I now move on to another key capability master information management in enterprise information management. Master information management (MIM) is comprised of the core business entities such as products, materials, cost centers, chart of accounts, customers, suppliers, business partners, employees, and geographic locations. MIM represents the high value information an enterprise uses repeatedly across numerous business processes, across multiple business units, and provides decisiveness to business processes.

Master information concerning customers, products, materials, and employees is scattered across a multitude of enterprise systems such as ERP, SCM, CRM, and HRMS. Each of these source systems has a variation of master information that becomes difficult to match and integrate across systems. Certain key data attributes may be missing, duplicate, or inconsistent across the systems holding master information. In this chapter the following topics are covered.

- *Define master information management*

- *The key drivers for master information management*

- *Building blocks for master information management*

- *Critical success factors for master information management*

- *Tools for master information management*

The next sections discuss each of these key considerations in more detail.

■ **Note**   The chapter goal is to explain what is master information management and the key drivers for master information management. Also how organizations can go about building the foundations for effective master information management, and the critical success factors to consider when embarking on a master information management program.

Although I introduced the concept of MIM, it is important to consider the supporting disciplines of information quality, metadata management, and semantics, which enhance the value the quality of master information in an enterprise. Also covered in this chapter is a real-world example of how an enterprise went about building a business case for an MIM program.

Definition of Master Information Management

Although there are numerous definitions of MIM, the most practical would be, to manage a single view of key master data entities, such as customer, product, chart of accounts, employees, suppliers, and so forth, of an enterprise that are created and consumed by multiple business units. Master information is crucial business information that supports both transactional and analytical operations of an enterprise. A single unified view of master data entities ensures that reporting is consistent across the enterprise, consistent in operational and supply chain execution, and has a common framework for enterprise planning and budgeting exercises. There are essentially three types of MIM—operational, analytical, and hybrid.

Operational MIM deals with a single view of master data information in the core systems of the enterprise where the master data is typically created and lack of a consistent view is bound to have an impact of the operational execution and efficiency of business processes. Here is an example—if the order management system and supply chain management have inconsistent views of the inventory of certain products, it will be certain that there will be defects in the operational efficiency of order management and execution. Analytical MIM deals with a single unified view of master data in the downstream system of record and analytical data marts that are largely consumed by business intelligence and analytical applications. Hybrid MIM involves handling master data in multiple ways, such as a combination of distributed data sources (aggregated through virtual MDM) and a single, golden record (physically stored in an MDM hub). This is often an enterprise's response to multiple use cases requiring different levels of information. As is evident from the different types of MDM, successful execution of MIM requires close cooperation of business and IT and should not be viewed as solely a technology solution.

Key Drivers for Master Information Management

The key drivers for MIM from an enterprise standpoint are as follows (see Figure 7-1):

- Growth
- Speed to market
- Cost optimization
- Enhance collaboration
- Single view of reporting
- Compliance

Figure 7-1. Business drivers for master information management

Growth

One of the key business drivers for MIM programs at the enterprise level is to promote growth of revenues through cross-sell and up-sell opportunities. To effectively grow, the customer share of wallet enterprises needs a 360 degree view of customer data, which is achieved through MIM. Effective MIM in the big data world also leads enterprises to build a customer view through both internal and external data (including social media data), which enhances the opportunities to tap high value customers for priority services.

Speed to Market

Consistent and governed master information processes ensure that customer onboarding time is less, thereby enhancing the customer acquisition process. Ability to integrate master information across business processes and business units in less time; enhances the ability to deliver customized offerings to customers based on preferences thereby enhancing the speed to market.

Cost Optimization

Single view of customer master data helps in reducing duplicates in customer master data resulting in cost optimization opportunities concerning elimination of multiple promotional mails to the same household. This helps the reduction in order processing issues due to inconsistencies in product inventory data.

Enhance Collaboration

Without a unified view of key master data, such as financial data across multiple sources, enterprises cannot share information effectively. Cross functional communication needs a common data language that everyone can use and to achieve this, an enterprise needs an analytical MIM in place. For example, a manufacturing company with multiple divisions involved in buying high quantities of the same product is unable to negotiate a bulk purchase contract resulting in an increase of cost materials. Hence the benefits of collaboration also drive down cost.

Single View of Reporting

With key master data, such as financial data scattered across multiple systems, many enterprises struggle to view one standardized set of financial metrics. Different systems produce different answers to the same question. Thus in the absence of a single view of master information, a single view of reporting becomes challenging. The problem is further complicated with different metrics meaning different things to different users. Therefore there is a need for metadata and business glossary definitions of metrics along with consistent master information to ensure a single view of reporting.

Compliance

Master information concerning customer opt-in and privacy preferences have a huge impact when driving customer reach out and campaign programs. Reaching out to customers who have opted out of contact programs can lead to legal and compliance issues with a reputation loss for the enterprise. Consistent master information also ensures accurate and consistent audit reports from a compliance standpoint.

A study of the value drivers gives a good starting point for enterprises and master information architects and business sponsors about how they should go about building their business case around MIM.

Building Blocks and Enablers for Master Information Management

With an understanding of MIM, I move on to the building blocks of MIM in an enterprise. Enterprises should build an MIM business case and a roadmap based on an assessment of the MIM capabilities in the enterprise and the targeted future capabilities based on an MIM vision. Figure 7-2 represents the phases of an MIM strategy. Table 7-1 defines the key activities and deliverables for each of the MIM assessment phases.

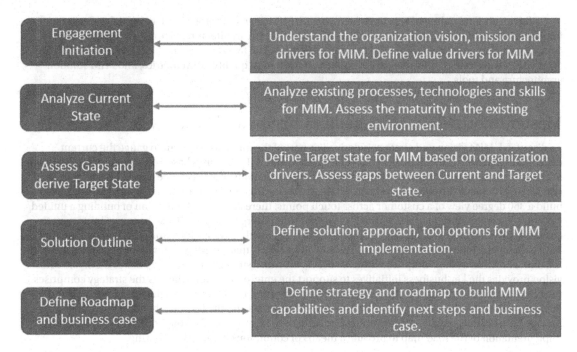

Figure 7-2. Master information management (MIM) assessment phases

Table 7-1. Activities and Deliverables for the Master Information Management (MIM) Assessment Phases

Engagement Phase	Activities and Deliverables
Engagement initiation	Engagement kickoff meeting with stakeholders. Define work plan and workshop schedules. Define the solution templates and project governance committee.
	Deliverables: Project plan, interview/workshop schedule, templates
Analyze current state	Develop MIM value driver trees and initial hypothesis. Map value drivers to hypothesis and current state solution architecture. Analyze current state issues and pain points. Conduct detailed interviews with stakeholders and subject matter experts.
	Deliverables: Current state assessment report
Assess gaps and derive target state	Conduct business capabilities workshops to derive enterprise MIM vision. Define capabilities to support the MIM vision. Define target state architecture and use cases to support business capabilities. Assess gaps to be addressed in the current processes and applications to achieve the target state. Deliverables: MIM business capabilities and target state architecture
Solution outline	Derive solution options. Perform tool evaluations and proof of concepts. Deliverables: Technology options with recommendations
Define road map and business case	Define the initiatives with business priorities. Define the implementation road map based on feedback from stakeholders. Define business case for first set of initiatives and define the next steps to implement these initiatives. Deliverables: Implement road map and MIM business case

With an understanding of how to get started with MIM at an enterprise level, I now delve deeper into the building blocks of MIM and discuss high-level solution architecture. Figure 7-3 shows the key components that go into building MIM into the enterprise. The five key building blocks to enable MIM are MIM vision and strategy, information governance, information quality, MIM metrics, and MIM solution architecture and tools.

Master Information Management Vision and Strategy

As discussed, MIM vision and strategy is derived as part of the initial assessment to gauge the current capabilities and future road map based on enterprise vision. MIM vision is based on the enterprise's mission or vision statement. It is important to understand how the benefits of MIM can be applied across the enterprise to achieve the enterprise vision. For instance, if a financial institution, such as a bank, wants to build a 360 degree view of a customer across touch points, there is a clear MDM vision of building a unified view of a customer master across channels to support this enterprise vision. It also becomes important to consider the value drivers of master information, such as growth, cost optimization, speed to market, that have a direct link to the enterprise vision. The key point here is that the MIM vision does not stand alone, but is a direct function of the enterprise vision. So MIM vision and strategy enables the enterprise vision; MIM vision provides the key business initiatives to support the enterprise vision whereas the strategy comprises the phased approach to achieve the vision. In close connection to the strategy is the MIM road map and business case that help to determine the sequence of initiatives and the business benefits the initiatives provide. The business case is defined upfront and needs to be monitored throughout the life cycle of the implementation of the road map to provide a means of continuous benefits monitoring.

Information Governance

As discussed in Chapter 6, MIM is one of the key disciplines covered in an information governance program. Governance of master information and a stewardship organization to support the creation and management of master data is crucial for the success of any MIM program. Information governance of master information includes defining policies to support business activities of an enterprise, to support compliance needs, such as the anti money-laundering aspects of the USA PATRIOT Act, customer data privacy needs as mandated by HIPAA, and so forth. The policies need to be enforced using an organization (data stewardship) that involve both business functions that create and consume the data as well as application system SMEs, which manage the applications that help manage the business process transactions.

Information Quality

Information quality is closely linked with the information governance of master information assets such as customer, product, suppliers, and so forth. Information quality ensures that decisions based on master information assets made by businesses can be trusted as the information quality is acceptable. Information quality also ensures that the information reported is consistent and can be used to report compliance to government acts and regulations (see Figure 7-3).

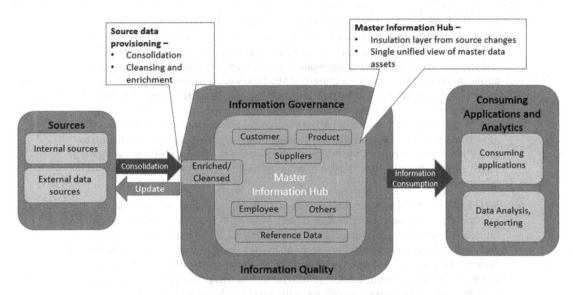

Figure 7-3. *Information governance and information quality in master information management*

To ensure that the information quality of the master information assets is of an acceptable standard the following activities need to be performed at the time of master information integration and the setting up of a master information hub.

- *Data profiling*—to assess the existing state of data quality. Data profiling to understand the duplicates in the master data or the gaps in linkages. Also, to understanding the scope of data enrichment to enhance the value of customer data assets. Poor information quality is one of the key drivers for MIM programs and hence data profiling is essential to ensure that quality of information in the master information hub is trustworthy. One key point to consider is that data profiling should not be seen only as a design time activity but as an ongoing activity to monitor the quality of information coming from source systems. See Figure 7-4 to understand the importance of data profiling and information quality.

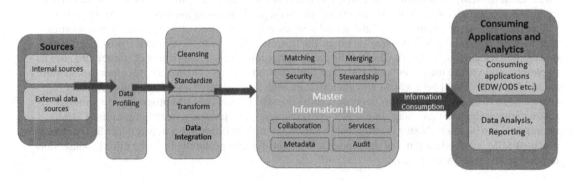

Figure 7-4. *Information quality in the master information management*

- *Data integration*—This is one of crucial design aspects of an MIM solution as integration of information into the hub happens through batch or real-time integration. Both batch and real-time integration can be handled through data integration tools, and real-time integration also can be handled through enterprise application integration tools. Commercial MIM tools also come with inbuilt data integration capabilities. The choice of the data integration solution should be based on the data volumes to be processed, the data latency needs, and the nature of the data to be integrated. In modern big data scenarios where unstructured data, such as web and social media data also are considered to be part of master information to give a 360 view of customer, the integration tools need to address such requirements. There also needs to be a decision on number of staging layers needed to process the information.

- *MIM hub*—The MIM hub can have a custom model or it can have a MIM tool purchased a with a customizable data model. Most MIM tools support multidomain implementations (e.g., product, customer, supplier, etc.). The hub needs to have audit capabilities to track the changes made to the system as well as integrate with the existing enterprise information security requirements. The design of the hub also needs to look at nonfunctional requirements such as performance, scalability, security, and maintenance capabilities of the MIM tool being considered.

- *Data matching, merging, and survivorship*—Master information integration process involves the elimination of duplicates. It also ensures that the surviving record has the best quality attributes. The MIM tool being considered needs to have fuzzy matching capabilities as well as merge capabilities to ensure that the attributes with latest data can be merged with the surviving record.

- *Synchronization of master information assets*—One of the key objectives of a master information solution is to provide master information assets of reliable quality to the right consumers at the right time. Hence the data synchronization design needs of MIM tools and solutions are required to keep in mind information consumption needs.

Master Information Management Metrics

The success of an MIM program at the enterprise level can be measured by the business benefits that the initiatives provide. However, it important to track the MIM program using metrics both as operational as well as business centric, which would give a bird's eye view to all business and IT stakeholders. Business metrics should be related to business outcomes. For instance, if customer retention is a business driver one could track customer retention before the implementation of the single view of customer and after its implementation to see the benefits of MIM. Cost optimization metrics concerning procurements spending should be implemented before MIM and after MIM as well. Business metrics always add significant confidence to management stakeholders concerning the effectiveness of MIM programs. Operational metrics concerning the percentage of duplicates in customer data, productivity benefits around MIM operational can be tracked using metrics. For instance, the time to introduce a new product SKU before MIM and post-MIM is a good measure of the time of market savings and productivity. I cover this in some more detail in a later real-life case study.

Master Information Management Solution Architecture and Tools

A key driver in any MIM is defining the solution architecture and supporting tools and technologies that enable the MIM processes. There are two key aspects in defining the solution architecture: 1) identify the architecture style to implement the system in any given enterprise and 2) identify the tools and technologies to build the solution. There are numerous architecture styles for implementing MIM. The common ones defined by Gartner are registry style, consolidation style, coexistence style, and transaction style.

> *Registry style*—In registry style implementation, the different source systems publish the master data, source system IDs, and any key data values used for matching to the subscribing hub. The hub runs its own cleansing and matching process to assign global identifiers to the matched master data records. The key point is that the golden copy of the master data is not stored in the hub and a 360 degree view is built by the hub in real time. The hub has an attribute location service that helps by finding the single version of truth. A registry style is suited for an enterprise where there is multiple source systems spread across the globe, and there is absence of authoritative systems of record.

> *Consolidation style*—The basic difference with the registry style and the consolidation style is that with the consolidation style, the hub stores a golden record of master data. The master data remains scattered over the source systems (spokes) and the master data is updated based on events but is usually not up-to-date. The master data in the hub is used more for reporting than for transactions. The hub cleanses, transforms, and matches data it integrates from source systems to build a golden copy of master data.

> Coexistence style—The coexistence style is similar to the consolidation style in that it has a golden record of the master data in the hub, which is sourced from the source systems (spokes). However the difference lies in the fact that in the coexistence style there is master data harmonization from which the master data is synchronized with the applications/source systems. However this approach is more time consuming and costlier.

> Transaction style—The transaction style is the most mature style where the master data is consistent, accurate, and up-to-date at all times. The key differences with the coexistence style are that all read and write operations for the master data are done in the hub. Any application that needs to be modified or created, the master data needs to invoke the MIM services of the hub. This ensures that the master data in the hub is up-to-date and there is absolute consistency in the master data.

With an appreciation of the master information implementation styles, there is a demand to drive the need for master information reference architecture in an enterprise. This involves deciding on the implementation style that is best suited for the unique needs of an enterprise and also deciding the method of implementation (operational/analytical or collaborative). Defining an MIM reference architecture also involves defining the key architecture principles depending on the information needs and maturity level of an enterprise. Some of key architecture principles would include the following:

- The MIM solution should provide the organization with a trusted, unified, and consistent view of master information. It should also control distribution of master information assets across the enterprise in a standardized manner.

- The MIM solution should be flexible to handle changes in business requirements, mergers and acquisitions, compliance needs, and the addition of new master information assets.

- The MIM solution should have the ability to decouple master information assets from enterprise applications and to make it available as a strategic asset for decision making.

As discussed earlier there is a need to define an MIM reference architecture. I cover two parts of the reference architecture—a logical architecture view of the MIM solution and the solution architecture view. The logical architecture view is shown in Figure 7-5. MIM logical architecture breaks up the reference architecture into the following building blocks: information providers and participants; services integration; information integration services; MIM services and information discovery such as visualization services, and metadata services.

- *Information providers and participants*—includes the internal and external participants that consume and update the master information assets as part of business processes and transactions as well as internal and external data providers (data sources). Internal participants could be business users as well as reporting and analytical applications. External data providers, such as Experian, Dun and Bradstreet, IMS, and AC Nielsen, also augment the customer, product sales data that is already present within an enterprise. External participants could be customers that access through various channels as well as business partners such as suppliers and vendors.

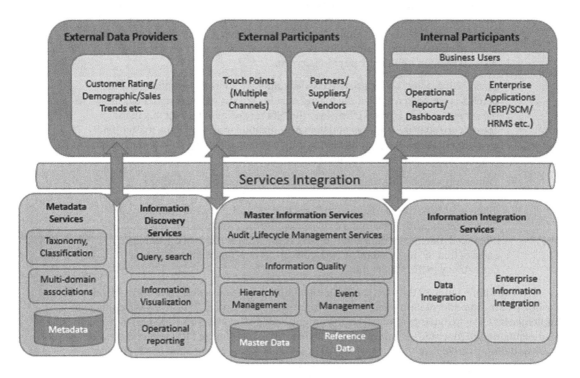

Figure 7-5. *Logical architecture of master information management*

- *Services integration*—ensures that the master information services and information integration services can be invoked from business applications without going through connectivity layers. Services integration is an enterprise services bus that provides integration between business applications and services.

- *Information integration services*—these include data integration services (for batch and near real-time integration) and enterprise information integration (for real-time integration needs).

- *MIM services*—these include the core MIM services including hierarchy management, event management, information quality, and audit and life cycle management services (which include the CRUD services and the associated business rules and logic).

- *Information discovery services*—These include visualization of master information assets stored in the master information repository. This becomes crucial as the move to a big data world where master information assets include both structured and semi-structured data. Discovery services also include query and search functionality of master information assets. In specific situations there is a need to have capabilities surrounding identity management as part of discovery services.

- *Metadata services*—These provide the capabilities to identify new categories of master information content and to create taxonomies to support the classification of content. Metadata services provide associated information about multidomain master information assets and their interrelationships.

With an understanding of the logical architecture for MIM, I address the solution architecture building blocks (see Figure 7-6).

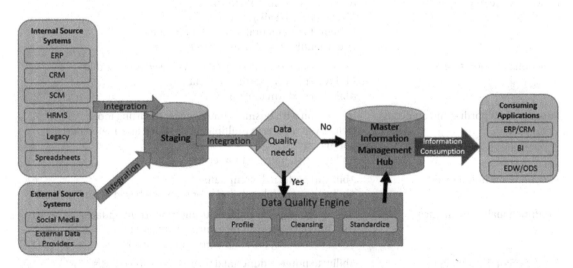

Figure 7-6. Solution architecture of master information management

The solution architecture building blocks include 1) a data profiling tool for data quality analysis and discovery; 2) data integration/enterprise application integration tools for integrating data in batch/real time from source systems and loading into staging area; 3) data cleansing capabilities to standardize, enhance, and cleanse the data; and 4) an MIM tool for providing the model; repository; and match, merge, and survivorship capabilities as well as the governance and workflow processes associated with managing the master information assets of an enterprise. The solution outline phase is when the tool evaluations are finished, and the required tools are selected. Here I briefly cover the process of MIM tool evaluation. There are a number of criteria on which MIM tools need to be evaluated. The evaluation criteria can be broadly classified into three areas namely, 1) functional capabilities, 2) nonfunctional capabilities, and 3) vendor capabilities. See Table 7-2 for MIM tool evaluation criteria.

Table 7-2. *Master Information Management Tool Evaluation Criteria*

Evaluation Area	Evaluation Criteria
Functional: Hierarchy management	Supports the creation and maintenance of hierarchies Supports attribute maintenance
Functional: Information governance	Supports governance processes for master information assets and change management Supports defining governance roles and privileges Supports for attribute based data security based on user roles
Functional: Workflow capabilities	Intuitive graphical user interface (GUI) for workflow design and customization Support for user workspace and version management Capability to integrate with external workflows
Functional: Search capabilities	Ability to perform drill downs Ability to search all attributes Support for configurable search parameters User friendly browsing interface for data
Functional: Information views	Ability to support customized views based on user role Ability to perform sorting on data Ability to perform tool tips
Functional: Reporting and analytics	Integration with business intelligence/reporting tools Delivery of reports in multiple modes such as e-mails/alerts/portals, etc. Ability to generate custom reports
Functional: Maintenance	Ability to perform bulk updates Ability to support batch and user interface based maintenance
Nonfunctional: Performance	User interface performance during concurrent updates User interface performance during batch loads Batch processing times for high volume bulk loads
Nonfunctional: Archiving	Ability to purge redundant data on a scheduled basis Ability to support archiving of master information assets
Nonfunctional: Security	Security mechanisms to control access to master information assets Ability to support single sign on and lightweight directory access protocol (LDAP) integration Ability to support encryption of master information assets

(continued)

Table 7-2. (*continued*)

Evaluation Area	Evaluation Criteria
Non Functional - Platforms and Databases	List of databases supported List of platforms supported Browser compatibility
Vendor - Product Support	Training support provided Quality of product documentation Support for new upgrades and patches Availability of migration path from older to newer product versions
Vendor - Country Specific Support	Vendor presence in the country of implementation Number of active clients and case studies Number of vendor consultants in the given country Number of consulting partners in the given country

Critical Success Factors in Master Information Management

Despite numerous efforts and investments in MIM there are numerous instances in which MIM programs have failed to deliver and produce the desired benefits. Although embarking on an MIM initiative with a maturity assessment to understand current capabilities and gaps is a good starting point, there is a need for a certain direction to sustain the momentum and realize the business benefits. In this section, the critical success factors in an MIM program are covered.

As discussed in the section about building blocks for MIM, there is a sequence of activities performed while embarking on an MIM program (see Figure 7-7).

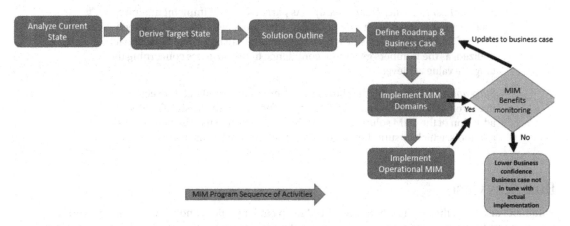

Figure 7-7. *Master information management program sequence*

The sequence of activities include analyze current state ➤ define target state and current gaps ➤ solution outline ➤ define road map and business case (initial) ➤ implement MIM domains that are defined based on the road map and monitor business benefits and feedback into the business case for a more realistic understanding of the business case ➤ implement operational MIM in which the master data changes are also fed back into the transaction/operational systems. The key critical success factors are as follows:

> *Build on a road map and business case*—Two of the key deliverables of the initial MIM strategy is the implementation road map (with a list of MIM initiatives to support the enterprise MIM vision) and a business case (qualitative or quantitative based on enterprise maturity and business buy-in). This is often the starting point of execution of an MIM program. (I cover a business case at the end of this section.)

> *Build incrementally*—Although there may be multidomain master information assets to be built as part of the MIM implementation road map, it is important to understand that MIM involves people, process, and technology and an incremental build is always a more pragmatic approach to deliver incremental business value. For instance, operational master information management is more complex to build as the changes/updates to master data need to be harmonized into the transaction/operational systems. Hence it is advisable to build a MIM hub first for analytical needs and then build operational MIM capabilities in a subsequent release.

> *MIM as part of information governance*—The need for information governance is crucial to make an MIM initiative succeed. Master information assets need to be governed. The processes that manage master information assets also need to be managed through enterprise governance teams. Ownership for the master information assets need to be defined.

> *Monitor business benefits continually*—As part of the MIM implementation as each of the initiatives are implemented, the business benefits achieved need to be fed into the initial business case so that there can be real-world feedback realized, as the numbers give more confidence to the business concerning the tangible value achieved.

> *Stakeholder buy-in*—MIM is a business driven exercise and needs executive buy-in as well as line-of-business (LOB) support for it to be successful. Greater user adoption of the MIM solution ensures that users learn to use the new processes and tools, which ensures better governance of the master information assets.

Business Case

Although I discussed the business benefits of MIM in an earlier section, I now dive into a real-world business case for a consumer good conglomerate that wanted to understand the impact of implementing an MIM solution for its fast moving consumer goods' (FMCG) business. The FMCG division had the following strategic business units (SBUs): incense sticks, cigarettes, matches, personal care products, foods, and cigarette exports. All master data for the FMCG division was stored in the FMCG SAP instance (each business unit had a separate SAP instance) including customer master, product master, logistics, and warehouse master data.

Master data organization: The FMCG team had a centralized team for managing the master data for all SBUs with the exception of foods, which had a separate team. The business case was based primarily around cost optimization and productivity. The focus was to understand the overall productivity benefits for the centralized MDM team and to understand productivity benefits with respect to product and materials master. The business processes for the generation of master data were studied and process enhancements were evaluated post implementation of the MDM tool. Existing process of product master data generation and proposed process are shown in Figure 7-8.

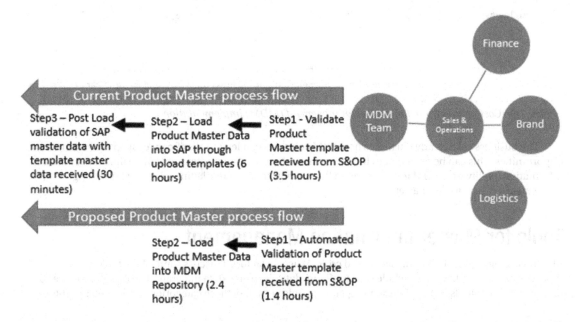

Figure 7-8. *Product master existing process and proposed process*

The study of these master data processes and the implementation of MIM revealed the following insights.

By automating the current Steps 1 and 2, which previously had been manual; Step 3 is automatically rendered redundant as the load process in Step 2 is automated. Productivity benefits of approximately 50% (worst case) to 60% (best case) can be achieved through improved efficiency of Steps 1 and 2 and the removal of Step 3.

An analysis of man days based on the study revealed the following findings shown in Figure 7-9.

MDM Team	Current Scenario – Man Days per month in managing master data (A)	Future Scenario – Man days per month with 50% savings (B)	Savings – Man months saved per annum (A-B)*12/22
Centralized MDM team	435.5	217.75	118.77
Foods MDM team	433.1	216.55	118.11
Total (All Teams)	868.6	434.30	236.88

No of man months saved in a year from MDM implementation
This when extended to other business units can lead to significant cost savings

Figure 7-9. *Cost savings from master data management (MDM) implementation*

The business case clearly illustrates how MIM solutions can lead to significant cost optimization opportunities. This can be extended to business cases where customer retention can be linked to enhanced information quality from MIM initiatives as well as savings in direct marketing costs by eliminating duplicates in the customer master.

Tools for Master Information Management

After reviewing the MIM building blocks and critical success factors, I discuss some of the tools in this domain. Some of the key tools include data profiling, data cleansing, data integration tools, MIM tools, and data visualization tools. Some of the leading market tools in each of these categories are covered in Table 7-3.

Table 7-3. *Tools for Master Information Management Solutions*

Information Quality	Informatica	IBM	SAP
Data profiling	Informatica Data Explorer	Infosphere Information Analyzer	SAP BusinessObjects Data Services
Data cleansing	Informatica Data Quality	Infosphere QualityStage	SAP BusinessObjects Data Services
Master information management	Informatica MDM	Infosphere MDM	SAP MDM
Data integration	Informatica PowerCenter	Infosphere DataStage	SAP BusinessObjects Data Services
Data visualization	Data Visualization in Data Archive is the closest match.	Watson Analytics	SAP Lumira

■ ■ ■

CHAPTER 8

Pillar No. 5: Information Warehousing

With an understanding of master information management and the key role it plays in managing master information assets in an enterprise, I now move to another important facility, information warehousing (traditionally known as data warehousing) in enterprise information management (EIM). Information warehouse is a consolidated system of records for all business information in an enterprise across business functions. It is a single version of truth that gives the business leaders a cross-functional view of enterprise performance. Building an information warehouse involves integrating data from numerous internal and in some cases external sources.

Business processes are executed and transactions performed through applications (e.g., ERP, CRM, SCM, HRMS, etc.) that generate data and are integrated into a single version of truth, which is called the system of record (SoR) or information warehouse. The primary purpose of the information warehouse is to provide a single version of truth for enterprise data for analysis and business intelligence purposes. In this chapter the following topics are covered:

- *Information warehousing definition*

- *The key drivers for information warehousing*

- *Building blocks for information warehousing*

- *Critical success factors for information warehousing*

- *Tools for information warehousing*

In the next few sections, I discuss each of these key considerations in more detail.

■ **Note** The chapter goal is to explain what information warehousing and key drivers are. Also how organizations can start building the foundations for an effective information warehouse, and the critical success factors to consider when embarking on an information warehouse program.

Although I have introduced the concept of an information warehouse, it is important to consider the supporting disciplines of data integration, data architecture, and metadata management that enhance the value of the information warehouse. Also covered in this chapter is a real-world example of how an enterprise went about building a business case for an information warehouse program.

Information Warehouse Definition

Although there are numerous definitions for an information warehouse, the most practical one would be that an *information warehouse* is a SoR that consolidates data from key enterprise applications as well as external data sources to give a 360 degree view of enterprise business. The SoR provides a data layer where all enterprise data relationships are captured, providing a robust data platform for performing cross-functional reporting as well as building function specific analytical data marts (finance data mart, sales data mart, etc.) One of the key drivers of an information warehouse is to provide a consistent, trustworthy data platform to enable business users to be self-sufficient when interacting with the data platform using a variety of business intelligence, analytical, and data visualization tools.

Some of the more conventional definitions include the classic Bill Inmon definition of enterprise data warehouse (EDW). A data warehouse is a subject-oriented, integrated, time-variant, and nonvolatile collection of data in support of management's decision-making process. Ralph Kimball another stalwart, describes an EDW as a collection of data marts with conformed dimensions using the data warehouse bus architecture. The two definitions also are closely related to two fundamental approaches to information warehousing namely the Bill Inmon or top-down approach and the Ralph Kimball or bottom-up approach. In the top-down approach a normalized EDW is designed and built first. Once the EDW is built, the dimensional data marts are built, which contain data related to specific business processes or specific business units. The EDW in the Inmon approach is also known as the SoR and serves as the single version of enterprise cross-functional data. The EDW resides at the center of the corporate information factory devised by Inmon (see Figure 8-1).

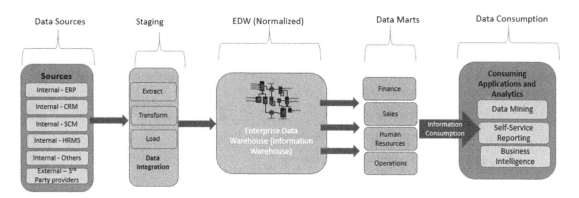

Figure 8-1. *Enterprise data warehouse (EDW) in the corporate information factory*

In the bottom-up (Ralph Kimball) approach the dimensional data marts are built first to facilitate reporting and analytics and then the data marts are combined through conformed dimensions to design and build the EDW. In the bottom-up approach the key business processes or business units are focused on building data marts and providing faster access to business intelligence and self-service reporting.

The data marts are designed as facts and dimensions with a focus on answering key business questions related to the business processes and business functions associated with the data marts. The focus of the dimensional design is to provide ease of use for business users as well as to answer queries. Once the data marts are designed, the EDW can be built by linking the data marts through a set of conformed dimensions in the data warehouse bus architecture. The Kimball or bottom-up approach can be represented by Figure 8-2.

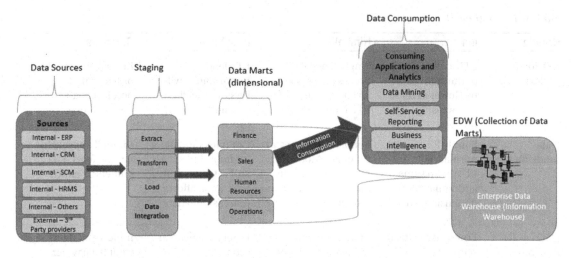

Figure 8-2. *Bottom-up approach to a data warehouse*

There also is a lesser-known third approach to building a data warehouse. This is called data vault and it was devised by Dan Linstedt. This is a hybrid approach that uses parts of the Inmon and Kimball methodology. The data vault is based on the concept of hubs, links, and satellites. Hubs are master tables with source system keys (e.g., customer, product, location, etc.). Links represent associations/relationships between hubs with a validity period for these relationships. Satellites point to links and contain detailed attributes of associated transactions and their period of validity. Data vault methodology also uses dimensional data marts to expose data to business users. Data vault models are used by some government agencies and where compliance needs are high. The key differences between the Inmon, Kimball, and data vault approaches are given in Table 8-1.

Table 8-1. *Differences Between Inmon, Kimball, and Data Vault Approaches*

Criteria	Inmon	Kimball	Data Vault	Comments
Time to market	Inmon approach takes time to build as first the normalized enterprise data warehousing (EDW) is built followed by dimensional data marts	Kimball approach ensures a quicker time to market as the data marts are built in an incremental manner and the EDW is essentially a collection of data marts.	Takes time to implement as dimensional data marts need to be built after the hubs, links, and satellites.	If time to market is a key differentiator, the Kimball approach provides the fastest business value.
Extract, transform, and load (ETL) complexity	More complex in Inmon as there are two sets of ETLs: source to staging to data warehouse, data warehouse to data marts.	Less complex; data is loaded directly to dimensional data marts. So only one set of ETLs is loaded from source to data marts.	More complex as the ETL process involves hubs, links, and satellite; table loads followed by dimensional data marts.	ETL is faster to build in the Kimball approach overall however transforms are complex as data mart is loaded directly.

(*continued*)

Table 8-1. *(continued)*

Criteria	Inmon	Kimball	Data Vault	Comments
ETL load performance	ETL load performance will be slower as there is a two-stage ETL process.	Single-stage ETL process results in faster execution of ETL batch process.	ETL load performance will be slower as there is a multistep ETL process.	Kimball approach provides the best ETL load performance.
Data modelling	Simple for normalized EDW with some additional effort for dimensional data marts.	Simple to design star schemas. Modelling effort is low as only data marts need to be modelled.	Complex modelling for hubs, links, and satellites with additional effort for dimensional data marts.	Kimball approach has the least data modelling effort.
Data integration needs	Enterprise wide in scope.	More department or business process driven.	Enterprise wide in scope.	Inmon and data vault have wider enterprise data needs.
Storage needs	Higher in Inmon approach as you need to store EDW data as well as data marts data.	Lower storage needs as only data mart data needs to be saved.	Higher in data vault as storage needs include hub, link, satellite, and data mart tables.	Kimball approach has the lowest storage needs.

There is no clear differentiator in terms of which methodology is adopted the most. Some key points to consider while making this decision include whether the data needs are enterprise wide in scope or driven more around specific business processes or departments, whether there are stringent compliance needs and audit requirements.

Key Drivers for Information Warehousing

The key drivers for information warehousing from an enterprise standpoint are as follows (see Figure 8-3):

- Single version of truth

- 360 degree view of enterprise performance

- Perform historical and time series analysis

- Trustworthy, consistent, and standardized information

- Platform for self-service business intelligence and analytics

- Productivity benefits

- Enhance a data lake for data discovery opportunities

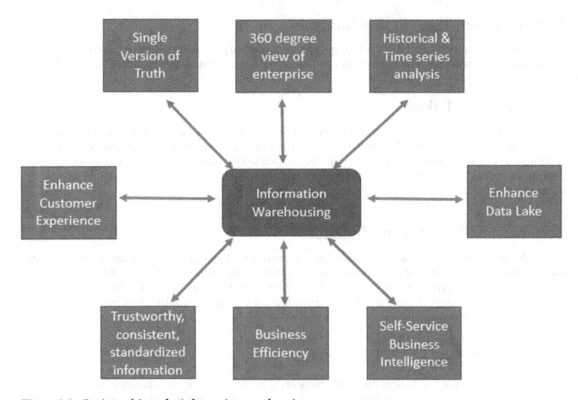

Figure 8-3. *Business drivers for information warehousing*

Enhance Customer Experience

One of the key business drivers for information warehousing is to provide a consolidated understanding of customer behavior across the different sales channels by integrating data from disparate sources. To effectively grow the customer share of business, enterprises need a 360 degree view of customer data, which is achieved through an information warehouse.

Single Version of Truth

Information integrated from disparate data stores provides a single version of enterprise data that can be leveraged for reporting and analytics. A single version of truth enhances the confidence of business decisions based on the metrics reported.

360 Degree View of the Enterprise

An information warehouse provides a 360 degree view of enterprise performance by integrating cross-functional data from disparate source systems. This gives enterprises an enterprise-wide view of business performance, which enables business decision making.

Historical and Time Series Analysis

Information warehouses store the entire history of transactions and master data, thereby providing the capabilities to perform trend analysis and time series analysis both of which need significant historical data.

Enhance Data Lakes

With the advent of big data solutions, a new concept of data lake has emerged. *Data lakes* are repositories of raw source data in their native format that are stored for extended periods. Although Hadoop based repositories are gaining momentum to store raw data as their storage costs are low, often data discovery use cases need structured reporting and analytics, which can be done with information warehousing thereby augmenting big data lakes.

Self-Service Business Intelligence

One of the key drivers of information warehousing is to provide business users with a scalable data platform for performing ad hoc reporting and analysis, thereby enhancing the self-service business intelligence capabilities of an enterprise. Self-service capabilities in turn drive the higher adoption of business intelligence tools as well as enhance business decision making through the availability of power data exploration and analysis capabilities.

Business Efficiency

An information warehouse enhances the productivity of the enterprise work force. Business users can quickly access critical data from a multitude of systems from a single repository and rapidly perform analysis and make smarter decisions. There is no need to spend time in collation of data from disparate sources, the focus is on analysis.

Trustworthy, Standardized, Consistent Information

The information warehouse provides a trustworthy, standardized, and consistent view of information extracted from a multitude of source systems. Data loaded into the information warehouse is profiled, standardized, and cleansed as part of the load process. This enhances the quality of data in the SoR and ensures business users that they can trust the data they consume.

Building Blocks and Enablers for Information Warehousing

With an understanding of information warehousing, I move on to the building blocks of an information warehouse in an enterprise. Enterprises should build an information warehouse business case and road map based on an assessment of information warehouse capabilities in the enterprise and targeted future capabilities based on the enterprise vision (see Figure 8-4).

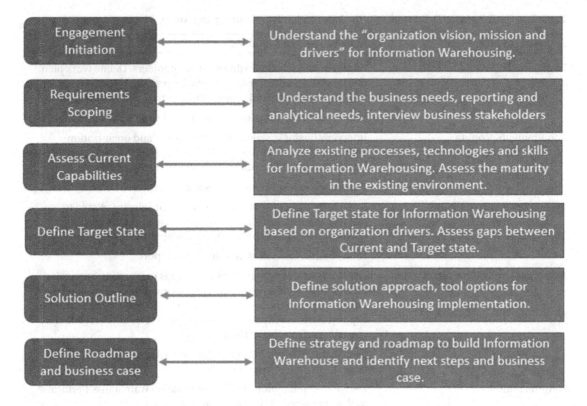

Figure 8-4. Phases of an information warehouse assessment

Table 8-2 defines the key activities and deliverables for each of the information warehousing assessment phases.

Table 8-2. *Activities and Deliverables for the Information Warehousing Assessment Phases*

Engagement Phase	Activities and Deliverables
Engagement initiation	Engagement kickoff meeting with stakeholders. Define work plan and workshop schedules. Define the assessment parameters, solution templates, and project governance committee.
	Deliverables: Project plan, interview/workshop schedule, templates
Requirements scoping	Interview business stakeholders to understand organization vision and business priorities. Understand the business priorities, business needs, reporting and analytical needs (measures, KPIs).
	Deliverables: Requirements scoping document
Assess current capabilities	Analyze existing processes, skills, and technologies related to information warehousing. Assess the maturity in the existing environment with respect to defined assessment parameters.
	Deliverables: Current state assessment report
Assess gaps and define target state	Conduct business capabilities workshops to ascertain enterprise information warehouse vision. Define capabilities to support the information warehouse vision. Define target state architecture to support business capabilities. Assess gaps to be addressed in the current state to achieve the target state.
	Deliverables: Information warehouse business capabilities and target state architecture document
Solution outline	Derive solution options for the information warehouse. Perform tool evaluations and proof of concepts.
	Deliverables: Technology options with recommendations
Define roadmap and business case	Define the information warehouse initiatives with business priorities. Define the implementation road map based on feedback from stakeholders. Define the business case for the first set of initiatives and define the next steps to implement these initiatives.
	Deliverables: Implementation roadmap and information warehouse business case

With an understanding of how to get started with information warehousing at the enterprise level, I now delve into the building blocks of an information warehouse. I also cover high-level solution architecture options later in this section. Figure 8-5 covers the key components that go into building an information warehouse. These include the following:

Data sources—the enterprise applications, operational systems that feed the information warehouse with source data on a predefined basis (batch or real time or near real time) depending on the business need for decision making.

Data integration/staging—The data integration tool extracts, transforms, and loads the source data into an information warehouse compliant form. In real-time integration needs, source systems changes are captured through the change data capture (CDC). The interim storing of the source data before loading into the SoR (information warehouse) is the staging layer that can be comprised of the initial staging and final staging. The initial staging is where the data is loaded directly from the source extracts and has a structure that is identical to the source extracts. The data is cleansed and transformed between initial staging and final staging. From final staging the data is loaded into the information warehouse. In certain use cases, such as operational reporting and near real-time reporting, there is a need for a separate repository, operational data store (ODS), which stores data in a normalized form for a limited period of 3 to 6 months. ODS can expose data through web services to near real-time reporting needs, such as in the telecom or banking industry to analyze fraudulent activities as transactions execute, or in mission critical systems such as oil field operations.

Information warehouse—The information warehouse is the SoR that is usually modelled as a third normal form (3NF). For enterprises that implement the Kimball approach this could be a dimensional model with a series of star schemas. The information warehouse retains the history of transactions and master data for as long as deemed necessary by the business. In a big data world with the existence of Hadoop based repositories in the enterprise, some of the older, less frequently accessed data could be archived from the information warehouse into the Hadoop repository.

Data marts—These are business processes or business unit specific data repositories that have subsets of information from the information warehouse. The data marts are used primarily for reporting and analytical purposes and are dimensional in nature.

Data consumption—This is the information consumption layer where the information warehouse/data mart data is consumed through a series of consuming applications (data mining/self-service reporting/business intelligence/operational reporting). In certain use cases involving real-time operational analytics and reporting the ODS data can be exposed through web services.

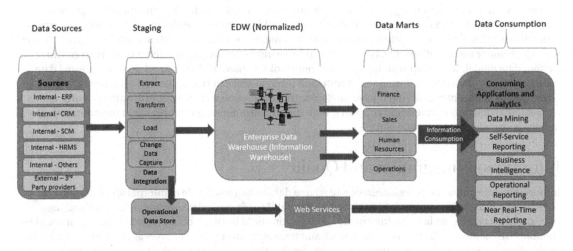

Figure 8-5. *The building blocks of an information warehouse*

The five key factors in the enablement of an information warehouse are information warehouse vision and strategy, information architecture, information integration and quality, information repository, and information warehouse solution architecture and tools.

Information Warehouse Vision and Strategy

As discussed earlier, information warehouse vision and strategy is derived as part of the initial assessment to gauge the current capabilities and future roadmap based on the enterprise vision and future business capabilities. Information warehouse vision is based on the enterprise's mission or vision statement and understanding how the benefits of the information warehouse can be applied across the enterprise to achieve the enterprise vision. For instance, if a financial institution such as a bank wants to build to improve customer experience across the different touch points, there is a clear information warehouse vision of building a unified view of customer transactions and feedback across channels to support this enterprise vision. It also becomes important to consider the value drivers of an information warehouse such as enhance customer experience, business efficiency, single version of truth, 360 degree view of enterprise performance that have a direct link to the enterprise vision. The key point here is that the information warehouse vision does not standalone but is a direct function of the enterprise vision. Therefore the information warehouse vision and strategy are enablers of the enterprise vision. The information warehouse vision provides the key business initiatives to support the enterprise vision whereas the strategy encompasses the phased approach to achieve the vision. In a close link to the strategy are the information warehouse road map and business case that help to determine the initiatives sequencing and the business benefits that those initiatives provide. The business case is defined up front and needs to be monitored throughout the implementation lifecycle to provide a means of measuring the real benefits.

Information Architecture

As discussed in the previous sections there are three ways to design an information warehouse, namely top-down (Inmon approach), bottom-up (Kimball approach), and data vault (Linstedt approach). One of the key points is to define the information architecture approach to be adopted for the information warehouse program. Information architecture includes defining the design approach to be adopted in the building of the information warehouse as well as deciding on the use of industry data models versus custom data models. Although the choice of industry models is driven by the enterprise's industry focus (retail/insurance/banking/energy, etc.) the bigger question enterprises often face is whether to go for an industry model or build a custom model from scratch. The key point to consider is the enterprise's business process's complexity and application maturity. Industry models are beneficial when leveraged out of the box with some minor modifications. In cases where extensive modifications are needed it is better to go with a custom data model. Information architecture not only looks at internal data sources but also external data sources and how information is exchanged with both internal and external systems. This is significant as large multinational enterprises often have multicountry information warehouses with data exchange needs with external business partners and suppliers. With the globalization of the supply chain the information architecture needs become more complex.

Information Integration and Quality

Information integration involves integration of source data from a multitude of internal and external sources. The integration approach is driven by the availability of source data, business decision-making needs, and accepted data latency. Information integration could be batch based or near real time. It could be standard ETL or in some instances extract, load, and transform. In certain use cases as in telecom or banking where there is a need to annualize transactions as they happen to detect fraudulent activities or provide customized offers to customers, the need for real-time data integration involves unique solutions such as streaming or the CDC based approach.

Information quality is very crucial as the information warehouse or SoR is considered the single version of truth for enterprise data. As there are numerous consumers of this enterprise scale, single version of truth, it is of paramount importance to ensure that the information quality of data that is integrated from the multitude of source systems (internal and external) is of acceptable quality. The key steps to information quality include data profiling of source extracts, data cleansing before loading data into the information warehouse, and data quality metrics to monitor the data quality in the information warehouse on a regular basis. Data profiling is the assessment of the existing state of data quality derived from the source systems. One important point to consider is that data profiling should not be considered only at design time, but as an ongoing activity to monitor quality of information integrated from source systems into the information warehouse. See Figure 8-6 to understand the importance of data profiling and information quality.

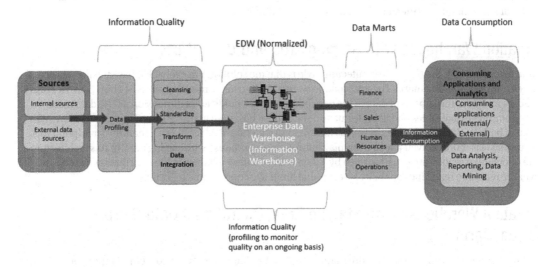

Figure 8-6. *Information quality in information warehouses*

Information Repository

Information repository is the actual store of enterprise data or the repository housing the information warehouse. There is another repository in the information warehouse architecture namely the staging database. The staging layer typically retains data for a short period of time (this can range from a few days to a month) depending on the business requirements.

Where there is a need for significant operational reporting and analytics, enterprises design a separate repository to store transactions over a period of three to six months. The reason for keeping a separate ODS is to ensure a repository that addresses specific operational queries and reporting needs and does not add the load of operational reporting on the information warehouse.

Information Warehouse Solution Architecture and Tools

A key in any information warehouse implementation is to define the solution architecture and supporting tools and technologies. There are two key aspects in defining the solution architecture 1) to identify the architecture style that implements the information warehouse for a given enterprise and 2) to identify the tools and technologies that builds the information warehouse solution. There are numerous architecture styles to implement an information warehouse. The common architectures include 1) an information warehouse with staging area; 2) an information warehouse with staging area and data marts; 3) an information warehouse with staging area, ODS, and data marts; and 4) an information warehouse with Hadoop repository (coexistence).

Information Warehouse with Staging Area

In this architecture style the data from source systems is integrated via a staging layer. The staging layer is usually comprised of two layers: the initial staging layer and the final staging layer. The source data is loaded to the initial staging layer, where all the data cleansing and data transformations are performed before they are loaded on to the final staging layer. In this architecture style the information warehouse is used for cross-functional reporting as well as operational reporting. A semantic layer also can be designed based on information warehouse tables to provide end users with a self-service business intelligence capability with minimal IT involvement. This architecture style is found in enterprises that have recently embarked on a journey to build information warehouses. In some situations the information warehouse may be modelled as dimensional to provide enhanced performance to reports. The information warehouse is treated as a collection of star schemas connected through conformed dimensions.

Information Warehouse with Staging Area and Data Marts

The basic difference with the previous architecture style is that the solution has evolved to have business process or business unit specific data marts. The information warehouse is usually modeled as 3NF (Third Normal Form) and the data marts are the subsets of data in the warehouse and are modeled as dimensional. The data marts also have some aggregate tables and the information warehouse has more granular data. The semantic layer can be designed both on data marts tables as well as information warehouse tables. From a maturity standpoint this architecture style is more evolved than a pure-play information warehouse with staging area. In this architecture style the operational reports are usually driven by a data mart or subset of tables in the information warehouse. However the real-time reporting capability is somewhat limited for operational reporting.

Information Warehouse with Staging Area, Operational Data Store, and Data Marts

The more complex reporting needs in more mature enterprises often require an additional repository called an ODS. The driver for a separate repository is twofold, the data granularity needs for operational reports as well as the latency and the time period of data needed is significantly different from the analytical and standard reports. This architecture style involves a staging area (initial and final staging) for feeding data into the information warehouse and downstream data marts. The ODS can be fed directly from the source systems through CDC (where data latency is a big factor) or normal ETL processes. The period of data storage in the ODSs can vary from a few weeks to six months depending on the operational reporting needs. This architecture style is usually found in more complex and mature enterprises with variable reporting needs.

Information Warehouse with Hadoop Repository (Coexistence)

With the advent of big data solution in the enterprise landscape there are new architecture patterns emerging where the positioning of the information warehouse is under discussion. With more unstructured data being integrated into the enterprise ecosystem and new data types, such as social media data, weblogs data, and machine data being analyzed, there is a need to store large volumes of both structured and unstructured data. Also the analysis patterns have changed and unstructured data is not required to be stored in structured repositories such as information warehouses. With the emergence of data lake repositories where large volumes of raw source data are stored in Hadoop repositories, and when there is a need to store these unstructured data they are converted into a structured format and stored in the information warehouse. Some examples of such data transfers are feedback or learning from customer campaigns may be stored in the information warehouse to keep a track of the effectiveness of campaigns. This results in a coexistence architecture where the information warehouse and Hadoop based repository coexist and augment each other (this is covered in greater detail in Chapter 11), see Figure 8-7.

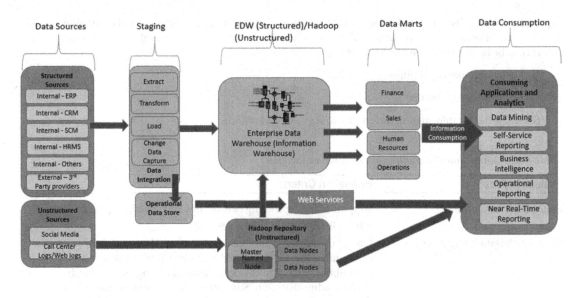

Figure 8-7. *Information warehouse coexistence with Hadoop repository*

Some key architecture principles surrounding information warehouses would include the following:

- The information warehouse solution should provide the organization with a trusted, unified, and consistent view of the cross-functional information. It should control the distribution of enterprise data assets across the enterprise in a standardized manner.

- The information warehouse solution should be flexible enough to handle changes in business requirements, mergers and acquisitions, compliance needs, and the addition of new subject areas.

- The information warehouse solution should be considered as the single version of truth and all decision-making applications must directly refer to the information assets of the warehouse or through the downstream data marts.

- In a big data enabled enterprise, the coexistence of the information warehouse and Hadoop should be seen as a means to augment the capabilities of each component rather than a replacement for either. The information warehouse and data marts are needed for use cases such as high-performance query zones while Hadoop is suitable for ETL processing workloads and storage.

As discussed in earlier pages of this chapter, there could be multiple architecture options for an enterprise that attempts to build an information warehouse. However the enterprise needs to define a target reference architecture based on those initiatives needed to ensure that the architecture vision is in line with the business vision. Figure 8-7, which shows the coexistence architecture between information warehouse and Hadoop, can serve as a reference solution architecture for mature enterprises.

The solution architecture building blocks include 1) a data profiling tool for data quality analysis and discovery; 2) data integration/change data capture tools for integrating data in batch/real time from source systems and loading into staging area; 3) data cleansing capabilities to standardize, enhance, and cleanse the data; 4) a database management tool to allow the repository to store the information warehouse data as well as the data in staging. The solution outline phase is after the tool evaluations are finished, and the required tools are selected. Here I briefly cover the process of information warehouse tool evaluation. There are a number of criteria based on which information warehouse tools need to be evaluated. The evaluation criteria can be broadly classified into two areas,

1. Functional capabilities

2. Vendor capabilities

See Table 8-3 for information warehouse tool evaluation criteria.

Table 8-3. *Information Warehouse Tool Evaluation Criteria*

Evaluation Area	Evaluation Criteria
Functional: Data profiling	Support for data profiling capabilities and visual/graphical representation of profiled data.
Functional: Data integration	Ease of use of tool graphical user interface, drag and drop capabilities, menus, shortcuts, and icons. Code versioning capabilities: check-in, checkout, time stamping. Support web-enabled administration. Support of diverse data sources, change data capture, complex transformations, workflow capabilities, handle large data volumes, parallelism of data loads, integration with Hadoop, metadata management, and audit trail capabilities.
Functional: Change data capture	Support for change data capture for diverse sets of sources and targets. Ability to handle large data volumes through change data capture.
Functional: Data cleansing	Ability to handle duplicates in source data, data enhancement capabilities. Ability to perform data standardization. Ability to define data quality rules for match, merge functionality. Ability to perform probabilistic matching. Ability to perform data linkages.
Functional: Database management	Ability to support diverse operations systems/hardware. Database scalability for data loading and concurrent usage. Scalability approach of the database. RAID (redundant array of independent disks) levels supported. Support for massively parallel processing architecture and parallel queries. Ability to perform under mixed query loads. Backup and archival mechanisms supported. Security features supported. Integration with portals/webpages. Integration with enterprise applications.
Vendor: Capabilities	Largest database instance/ETL instance Different licensing modes and costs Collaborating with hardware and software vendors
Vendor: Product support	Training support provided Quality of product documentation Support for new upgrades and patches Availability of migration path from older to newer product versions
Vendor: Country specific support	Vendor presence in the country of implementation Number of active clients and case studies Number of vendor consultants in the given country Number of consulting partners in the given country

Critical Success Factors in Information Warehousing

Despite numerous efforts and investments in information warehouse initiatives there are numerous instances where information warehouse programs have failed to deliver and produce the desired benefits. Although embarking on an information warehousing initiative with a maturity assessment to understand current capabilities and gaps is a good starting point, there is a need for a certain direction to sustain the momentum and realize the business benefits. In this section, the critical success factors in implementing an information warehouse are covered.

As discussed in the section on building blocks, there is a sequence of activities performed while embarking on an information warehouse program, see Figure 8-8.

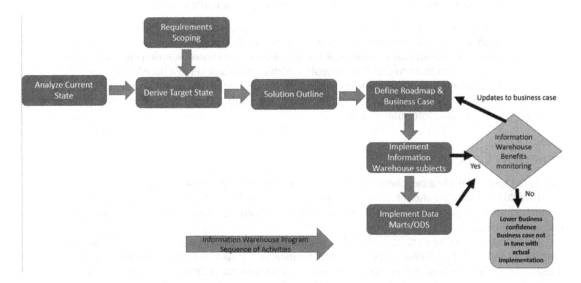

Figure 8-8. *Information warehouse program sequence*

The sequence of activities include analyze current state and requirements scoping ➤ define target state and current gaps ➤ solution outline ➤ define road map and business case (initial) ➤ implement information warehouse subjects that are defined based on the roadmap and monitor business benefits and feedback into the business case for a more realistic understanding of the business case ➤ implement data marts and ODS based on the enterprise needs. The key critical success factors are as follows:

> Build on road map and business case—Key deliverables of the initial information warehouse strategy are the implementation roadmap (with a list of information warehouse initiatives to support the enterprise business vision) and a business case (qualitative or quantitative based on enterprise maturity and business buy-in). This is often the starting point of the execution of an information warehousing program. (I cover a business case at the end of this section.)

> Build incrementally—Although there may be multidomain subject areas to be built as part of the information warehouse implementation road map, it is important to understand that the building involves people, process, and technology, and an incremental build is always a more pragmatic approach to deliver incremental business value. For instance, data marts/ODS are more complex to build in the absence of a SoR (information warehouse). Therefore it is advisable to build an information warehouse first for analytical needs and then build ODS capabilities in a subsequent release.

Information warehouse as part of information governance—The need for information governance is crucial to make an information warehouse initiative succeed. Information warehouse assets will need to be governed. Data exchange processes that manage the exchange of information assets both within and outside the enterprise also need to be managed through enterprise governance teams. Ownership of the information warehouse assets needs to be defined.

Monitor business benefits continually—As each of the initiatives become implemented (as part of the information warehouse implementation), the business benefits achieved need to be fed back into the initial business case for real-world feedback. The feedback gives credibility to the numbers, which in turn gives more confidence to the business concerning the tangible value achieved.

Stakeholder buy-in—Information warehouse is a business driven exercise and needs executive buy-in as well as line-of-business (LOB) support for it to be successful. Greater user adoption of the information warehouse solution ensures that the users learn to use the new tools, which ensures self-service based information consumption, productivity benefits, and more time spent on actual analysis and decision making as compared to data collation and governance of the enterprise information assets.

Business Case

Although I spoke of the business benefits of information warehouse in an earlier section, I now dive into a real-world business case and return on investment (ROI) model for a media conglomerate that wanted to understand the impact of implementing an information warehouse solution for its newspaper and magazines business unit.

Case study: A media company was looking to understand the potential business benefits for its advertising and sales, circulation and distribution, finance, newspaper, and magazine brands for an information warehouse implementation. The business requirements to drive these business benefits were defined through a set of meetings and workshops. The benefits for a quantified and a ROI model were generated, and an implementation roadmap was defined to take the benefits to fruition. Figure 8-9 shows the current challenges that can be converted into business benefits by the implementation of the information warehouse.

Circulation and Distribution

Current Challenges	Benefits
• Ability to track & analyze Advertising Revenues by region, brand wise revenue analysis.	• Enhance Revenue Potential through Advertising Revenue Analysis
• Seamlessly track kind of advertisements that cover spaces such as scheme adds, offers, new launches, etc.	• Discover Industry Trends in Advertising Types Analysis
• Ability to analyse industry trends from external data sources like TAM, MAP, IRS, NRS, etc.	• Discover Industry trends through Information Integration of External Data Sources
• Ability to perform ad hoc reporting of circulation data, analysis of unsold from Agents, impact of campaigns on circulation	• Insights into Unsold, Campaigns through Circulation Analytics
• Ability to have an integrated view of business across standalone reports of production, circulation, distribution, finance, ability to integrate circulation sales data with add revenues and readership data	• Integrated View of Business Operations through Integrated Value Chain Analysis & Reporting

Figure 8-9. *Challenges converted to benefits for circulation and distribution business unit*

This exercise was repeated for the other business units and the benefits were quantified wherever possible in discussion with the business unit owners. The benefits were classified into three heads—productivity benefits, customer retention benefits, and cost optimization benefits.

The benefits were calculated over a year and these were prorated for a three-year period as part of building the cost benefits ROI model. Figure 8-10 gives a representation of the information warehouse implementation business benefits.

Business Benefit Area	Benefit in Millions over 3 years	% Contribution
Cost Savings	1.6	9.14
Productivity	3.6	20.57
Customer Retention	12.3	70.29
Total Benefits	17.5	100

Benefits in Millions over 3 years

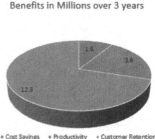

■ Cost Savings ■ Productivity ■ Customer Retention

Figure 8-10. *Information warehouse business benefits*

The implementation costs were developed using a combination of software options to obtain the most optimal one. The implementation costs were based on software costs, hardware, consulting, personnel, training, and the depreciation of software and hardware.

Based on the costs and benefits the ROI model was developed to obtain the net present value (NPV) and payback period. These are important measures to consider while quantifying the business case for an information warehouse implementation. In certain situations it becomes difficult to quantify the business benefits in which case the best case and worst case scenarios can be considered and later the business case can be modified based on actual benefits after the first set of initiatives are implemented.

Tools for Information Warehousing

After reviewing the information warehousing building blocks and critical success factors, I discuss some of the tools in this domain. Some of the key tools include data profiling, data cleansing, data integration tools, database management tools, and CDC tools. In the big data coexistence architecture there are other tools involved as well, however, these are covered in detail in Chapter 11. Table 8-4 covers some of the leading market tools in each of the categories.

Table 8-4. *Tools for Information Warehousing Solution*

Information Quality	Informatica	IBM	SAP
Data profiling	Informatica Data Explorer	Infosphere Information Analyzer	SAP BusinessObjects Data Services
Data cleansing	Informatica DataQuality	Infosphere QualityStage	SAP BusinessObjects Data Services
Database management	No capability	Pure Data for Analytics, Pure Data for Operational Analytics	Sybase IQ, SAP HANA
Data integration	Informatica PowerCenter	Infosphere DataStage	SAP BusinessObjects Data Services
Change data capture	Informatica CDC	Infosphere Data Replication	SAP Sybase Replication Server

■ ■ ■

CHAPTER 9

Pillar No. 6: Information Delivery and Consumption

With an understanding of information warehousing and the key role it plays in providing a single version of truth to an enterprise, I now move to another key capability information delivery and consumption (traditionally known as business intelligence [BI], reporting, and the rendering of it through various channels) in enterprise information management (EIM). Information delivery and consumption is the last component in the EIM space and deals with the consumption of structured and unstructured information through a series of channels such as web browsers, portals, web services, and mobile devices. Although I discussed how information can be processed, stored optimally, and can be trustworthy to end consumers, its ultimate value comes from how it is delivered, and that is timely and through the optimal channels to key decision makers. To understand the information consumption needs of an enterprise it is crucial to design an information delivery and consumption framework that addresses the needs of information at different levels and in-line with the enterprise vision and goals.

The primary purpose of the information delivery and consumption layer is to provide a timely and consistent view for enterprise data through the right channels for analysis and BI purposes. In this chapter the following topics are covered:

- *Information delivery and consumption definition*

- *The key drivers for information delivery and consumption*

- *Building blocks for information delivery and consumption*

- *Critical success factors for information delivery and consumption*

- *Information security challenges concerning information delivery and consumption*

- *Tools for information delivery and consumption*

The following sections discuss each of these key considerations in more detail.

■ **Note**   In this chapter my goal is to explain information delivery and consumption and the key drivers for information delivery and consumption. I also discuss how organizations can build foundations for an effective information delivery and consumption layer and the critical success factors to consider when embarking on information delivery and consumption related initiatives. There is also a discussion concerning information security challenges when dealing with information delivery and consumption.

Although I introduced the concept of information delivery and consumption, it also is important to consider the supporting disciplines of performance management, information governance, and metadata management that enhances the significance and value of the information delivered through information delivery. I also provide a real-world example of how an enterprise built a performance management framework (see Figure 9-6) through an information delivery and consumption strategy.

Information Delivery and Consumption Definition

Although there are numerous definitions for information delivery and consumption (commonly known as BI), the most practical one would be—Information delivery and consumption is a decision support system that consolidates discrete pieces of information from enterprise applications and other sources and provides management, operations managers, and administrative staff with a consolidated, consistent set of enterprise financial, operations, and performance metrics that enable decision making.

Some of the more conventional definitions include the classic Gartner definition of business intelligence—BI is an umbrella term that includes applications, infrastructure, and tools, and best practices that enable access to and analysis of information to improve and optimize decisions and performance. The goal of BI is to enhance the interpretation of large volumes of enterprise data. The ability to identify new opportunities based on data analysis and to implement the strategy to support these opportunities can serve as a market differentiator.

Closely related to information delivery and consumption is the concept of performance management. There are numerous methods to measure the performance of an enterprise. The core of performance management is measuring the effectiveness of a business process or business unit through a set of key performance indicators. One of the most widely adopted performance management frameworks is the Balanced Scorecard technique designed by Kaplan and Norton. Balanced Scorecards involve looking at an enterprise (private, public, or nonprofit) through four perspectives, namely—financial, customer, learning and growth and operations. The Balanced Scorecard framework can be adopted across multiple industries and in some instances there may be additional perspectives such as health, safety, and environment in some industries (as in the case of the oil and gas and mining industries). The relationship between BI and performance management lies in the fact that BI systems are decision support systems that help enterprises measure performance and also analyze the effect of execution with respect to targets and overall enterprise vision. Performance management frameworks help in providing the key perspectives of performance and the associated metrics that need to be monitored and analyzed. BI tools serve as enablers of the performance management frameworks to monitor the key result areas of an enterprise. Some of the other performance management frameworks like Six Sigma are also used to measure and improve manufacturing processes. Six Sigma originated at Motorola as a set of practices, designed to improve manufacturing processes and eliminate defects. Over time it has been adopted in a wide variety of industries including software services.

Although BI has traditionally been used to support decision making, with the advent of performance management, the maturity of the solutions has been expanded to include the effectiveness of execution to an enterprise strategy. See Figure 9-1 to help understand the relationship between the two.

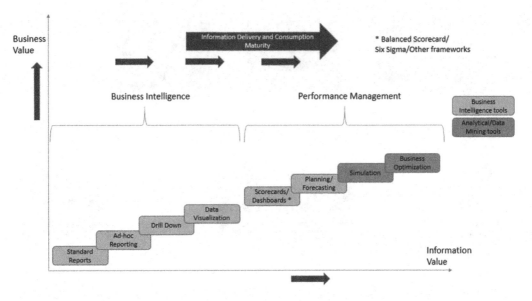

Figure 9-1. *Information delivery and consumption and performance management*

Some of the other common applications of information delivery and consumption include business activity monitoring. The key drivers for information delivery and consumption are covered in the next section.

Key Drivers for Information Delivery and Consumption

The key drivers for information delivery and consumption from an enterprise standpoint are as follows (see Figure 9-2):

- Self-service business intelligence
- 360 degree view of enterprise performance
- Perform historical and time series analysis
- Enhance customer experience
- Productivity benefits
- Supply chain optimization
- Manufacturing efficiencies
- Operational Intelligence

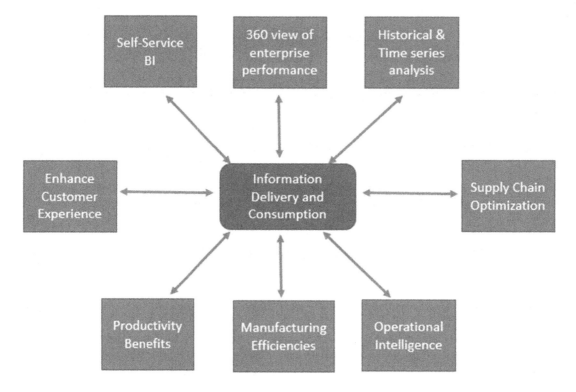

Figure 9-2. *Business drivers for information delivery and consumption*

Self-Service Business Intelligence

One of the key drivers of information delivery and consumption is to provide business users with a semantic data layer for performing ad hoc reporting and analysis, thereby enhancing the self-service BI capabilities of an enterprise. Self-service capabilities in turn drive the higher adoption of BI tools as well as enhance business decision making through the availability of data exploration and analysis capabilities.

360 Degree View of the Enterprise

This provides a 360 degree view of enterprise performance by the integration of cross-functional data from disparate source systems. It also gives them an enterprise-wide view of their business performance, which in turn enables business decision making. Balanced Scorecards based on cross-functional metrics can provide unique insights into business performance and measure efficiencies in business execution.

Perform Historical and Time Series Analysis

Information delivery and consumption provide a semantic layer that accesses the information warehouse. The information warehouse stores the entire history of transaction and master data, thereby providing the capability to the enterprise to perform trend analysis and time series analysis both of which need significant historical data.

Enhance Customer Experience

One of the key business drivers for information delivery and consumption is to provide a consolidated understanding of customer behavior across the different sales channels by analyzing integrated customer data from disparate sources. The enhanced customer experience can be achieved through a three-stage process: identify ➤ understand customer behavior ➤ interact.

Identify involves understanding the customer profitability and customer lifetime value. Understand involves developing the customer segments, buying behavior, and churn trends. Interact is the action stage where targeted campaigns are executed, campaign analysis is performed, and cross selling is done. Information delivery and consumption play a vital role in conjunction with customer relationship management (CRM) applications at each stage of this process to enhance the customer experience.

Supply Chain Optimization

With the advent of global supply chains for large transnational enterprises, managing supply chains is a huge challenge. Information consumption and delivery ensures that consolidated supply chain data from manufacturing to order execution, with a 360 degree view to supply chain planning and execution. Supply chains can be further optimized based on the consolidated planning, operations, and execution data to reduce costs concerning inventory, distribution, logistics, and so forth.

Productivity Benefits

An information delivery and consumption layer enhances the productivity of the enterprise work force. Business users can quickly access critical data from a multitude of systems integrated through a semantic layer and rapidly perform analysis and make smarter decisions. There is no wasted time spent in collation of data from disparate sources, the focus is on analysis.

Manufacturing Efficiencies

The delivery and consumption layer helps manufacturing companies optimize their inventory to reduce overcapacity as well as ensure sufficient supplies. Financial management in terms of operating expenses, cash-to-cash cycle time also help manufacturers focus on both profit as well as cost optimization opportunities.

Operational Intelligence

More enterprises are building capabilities in operational BI to provide insights into bottlenecks and efficiencies of business processes. There are two key characteristics of operational BI: 1) operational intelligence in an operational data store/information warehouse ecosystem and 2) operational intelligence embedded in business events and processes (analysis of events as they happen). The benefits are manifold including better and faster decision making; actionable insights into business operations and processes; and enhanced performance through automation, lower costs, and the ability to integrate business operations with enterprise strategy.

Building Blocks and Enablers for Information Delivery and Consumption

With an understanding of information delivery and consumption, I move on to the building blocks of information delivery in an enterprise. Enterprises should build an information delivery and consumption business case and road map based on the assessed performance management and business analytics needs in the enterprise, and targeted future capabilities based on enterprise vision (see Figure 9-3).

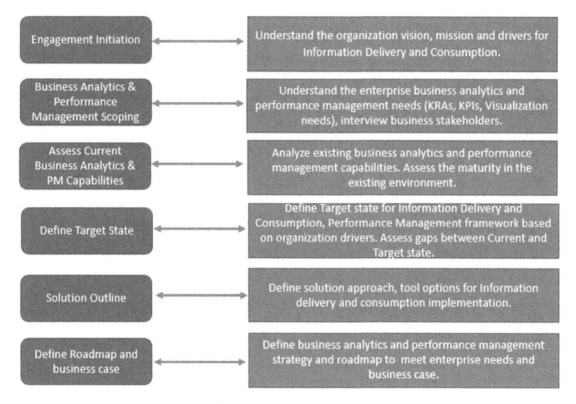

Figure 9-3. *Phases of an information delivery and consumption strategy*

Table 9-1 defines the key activities and deliverables for each of the information delivery and consumption strategy phases.

Table 9-1. *Activities and Deliverables for the Information Delivery and Consumption Strategy Phases*

Engagement Phase	Activities and Deliverables
Engagement initiation	Engagement kickoff meeting with stakeholders. Define work plan and workshop schedules. Define the assessment parameters, solution templates and project governance committee.
	Deliverables: Project plan, interview/workshop schedule, templates
Business analytics and performance management scoping	Interview business stakeholders to understand organization vision, business analytics, and reporting priorities; and information visualization needs. Understand the performance management drivers, supported frameworks, key result areas with KPIs (key performance indicators), metrics.
	Deliverables: Business analytics and performance management scoping document
Assess current business analytics and performance management capabilities	Analyze existing processes, skills, and technologies related to business analytics and performance management. Assess the maturity in the existing environment with respect to defined assessment parameters.
	Deliverables: Current state business analytics and performance management assessment report.
Define target state	Conduct business capabilities workshops to derive enterprise business analytics and performance management vision. Define capabilities to support the business analytics and performance management vision. Define target state architecture to support the required capabilities. Assess gaps to be addressed in the current state to achieve the target state.
	Deliverables: Information delivery and consumption business capabilities and target state architecture document
Solution outline	Derive solution options for the information delivery and consumption solution. Perform tool evaluations and proof of concepts.
	Deliverables: Technology options with recommendations
Define road map and business case	Define the information delivery and consumption initiatives with business priorities. Define the implementation road map based on feedback from stakeholders. Define business case for the first set of initiatives and define the next steps to implement these initiatives.
	Deliverables: Implementation road map and business case

With an understanding of how to get started with information delivery and consumption at an enterprise level, I now delve into the building blocks of an enterprise business analytics and performance management solution. High-level solution architecture options are also covered later in this section. Figure 9-4 covers the key components. These include the following:

Information warehouse, operational data stores—the decision support systems, such as the information warehouse, data marts, and operational data stores, serve as data sources for the consolidated, consistent view of enterprise information for information delivery and consumption.

Semantic layer—In some instances, a semantic layer is built to provide power users and business analytics a self-service reporting capability. The semantic layer is a virtual layer over all data sources (information warehouse, data marts, ODS) and presents the underlying data subjects from a business standpoint for easier readability and comprehension.

Information delivery and consumption—The information derived from the decision support systems or semantic layer is delivered to numerous business analytics, data mining, and performance management applications. Some of these applications include—data mining, self-service reporting, planning and budgeting, scorecards and dashboards, business activity monitoring, and near real-time operational reporting. There are numerous channels of information consumption including web browsers, enterprise portals, web services, and mobile devices. The design of any information delivery and consumption solution needs to factor in the information consumption channels while defining the solution architecture and design.

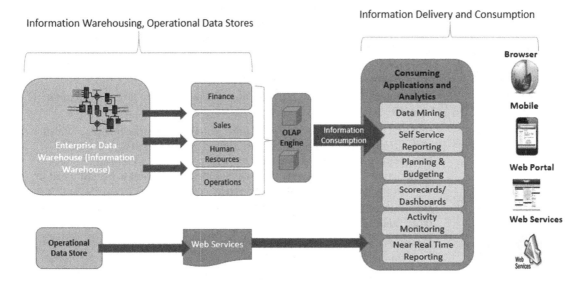

Figure 9-4. *The building blocks of information delivery and consumption layer*

I now give a brief description of each of the information consumption applications.

- *Data mining*—These are applications that mine the enterprise data to detect hidden patterns around customer behavior, supply chain efficiencies, and employee behavior to enable the enterprise to make smarter decisions. The data mining algorithms run on both structured and unstructured data depending on the use cases in question.

- *Self-service reporting*—This enables business analysts and power users to analyze the enterprise data and ask ad hoc questions that help discover nuances about the enterprise data. Self-service reporting is one of the key applications from an information delivery and consumption perspective.

- *Planning and budgeting*—As enterprises deal with dynamic planning scenarios there is a greater need for planning and budgeting applications that can handle dynamic planning scenarios due to unforeseen market conditions and also enable faster planning and decision-making cycles as compared to traditional planning processes. This is one of the key performance management applications in the information delivery and consumption layer.

- *Activity monitoring*—Enterprises often need to analyze and report on events and alerts in business processes using real-time dashboards and event-based alerts. This is often crucial in certain automated businesses that may need manual intervention when the alerts are being generated based on some threshold value that has been exceeded.

- *Scorecards and dashboards*—Scorecards are used by an enterprise to measure the progress against the enterprise strategy. Scorecards represent performance trends over a period of time, such as month/quarter/year, whereas dashboards indicate the status of a performance metric at a given point in time. Scorecards are used to display performance management frameworks such as Balanced Scorecard, and Six Sigma. Scorecards are defined based on key performance indicators of the business and are based on the performance management framework that the enterprise chooses to follow. As scorecards mostly measure lagging measures, the data latency needs are not high and real-time information is not required. Scorecards also are comprised of summarized metrics and ratios and are not the granular data one finds in reports and dashboards. The power of scorecards also is enhanced by data visualization enhancers such as dials, slider bars, and graphs. Dashboards in contrast are used to represent actual granular data, they contain data that is more recent that scorecards. Dashboards also cover performance metrics, but the grain of information is more detailed than the trends displayed in a scorecard.

- *Near real-time reporting*—Operational reporting in near real time to support operational BI needs is the requisite of operations-focused businesses. Oil rigs that run 24/7 often need operational reporting of key rig performance parameters in near real time, which are monitored closely by the operations team. Some other examples include stock market indexes and stock value monitoring.

The four key factors in the enablement of an information delivery and consumption layer are information delivery and consumption vision and strategy, information delivery and consumption self-service capabilities, information consumption channels, and information delivery consumption solution architecture and tools.

Information Delivery Consumption Vision and Strategy

As I discussed, information delivery and consumption vision and strategy is created as part of the initial assessment to gauge the current capabilities and future road map, which was based on the enterprise vision, future analytics, and performance management needs. Information delivery and consumption vision is based on enterprise's mission or vision statement, and understanding how business analytics and performance management can help in realizing that vision. The key point here is that information delivery and consumption vision is not a standalone strategy but a direct function of the enterprise vision. Information delivery and consumption strategy defines the business analytics and performance management initiatives that help an enterprise to achieve its vision. In a close link to that strategy are the information delivery and consumption road map and business case that help to determine the sequence of initiatives and the business benefits the initiatives provided. The business case is defined upfront and needs to be monitored throughout the implementation life cycle to provide a means of continuous benefits monitoring.

Information Delivery and Consumption Self-Service Reporting Capabilities

As discussed in the previous section, one of the key drivers for information delivery and consumption is to enhance the self-service reporting capabilities of an enterprise. One of the key success factors in an information delivery and consumption program is to ensure a wider adoption of business analytics and performance management applications by business users and line managers. Hence one of the key initiatives is to explore and enhance self-service reporting capabilities by building a semantic layer and ad hoc reporting capabilities to power users. It is also important for business analysts to explore patterns in the data by asking ad hoc questions.

Information Delivery and Consumption Channels

Enterprises are increasingly going global and with a mobile workforce, one of the key challenges is to keep executives and managers informed of business events and the changing conditions while on the move. Information delivery not only needs to be secure, consistent, and timely but also must be delivered through appropriate channels. For sales force automation the need for mobile based BI and reporting is a key requirement. Also with closer relationships with suppliers and vendors in a global supply chain, enterprise portals are increasingly used to provide inventory and order data to suppliers and vendors. These trends have resulted in a profusion of information consumption channels such as web browsers (thin clients), mobile, supplier and customer portals, and web-based services integrated with business partners and external agencies. In certain industries such as banking and financial services, they need to report their financial numbers through data submission applications based on XBRL (extensible business reportable language) for compliance to central bank regulations.

Information Delivery and Consumption Solution Architecture and Tools

A key driver in any information delivery and consumption solution is to define the solution architecture and supporting tools and technologies. There are two key aspects in defining solution architecture: 1) identify the solution components needed to build business analytics and performance management capabilities in a given enterprise, and 2) identify the tools and technologies needed to build the information delivery and consumption solution. The key solution components that go into building an information delivery and consumption solution are as follows: 1) BI tool for static reporting, mobile BI capabilities, scorecards, and dashboards; 2) business activity monitoring tool for event-based alerts and reporting; 3) planning and budgeting tool; 4) data visualization tool for specific data visualization needs; 4) business glossary based tool for a view of business metadata to business users; 5) data mining tools; and 6) enterprise portal tools for providing portal based access.

> Business intelligence tools—BI tools provide some key capabilities to the solution concerning static reporting, semantic layer, ad hoc reporting/self-service functionality, scorecarding and dashboarding, and mobile BI capabilities. However in some instances separate mobile applications need to be developed in certain use cases in which the mobility solution not only needs to read data but also write back data into the information delivery solution based on customer feedback or sales force notes.

Business activity monitoring tools—In certain types of business functions, there is a need for event-based reporting or alert-based notifications. This requires an events logging capability that is not present in standard off-the-shelf BI tools and calls for a *business activity monitoring tool* (a term first used by Gartner). Business activity monitoring (BAM) tools are used to provide a real-time summary of business activities and events to operations managers and decision makers. One key difference between BI dashboards and BAM dashboards is the data latency—BI dashboards report data refreshed at a predefined interval, where BAM dashboards report on near real-time data.

Planning and budgeting tools—Planning, budgeting, and forecasting tools provide the enterprise with capabilities to build plans that define the financial directions and goals for a period of three to five years. They also help to define budgets concerning how the plan will be executed month by month and quarter by quarter with an emphasis on expenditure. Forecasts also are defined based on historical data, business events, and seasonal variations to predict financial outcomes for a given period of months and/or years.

Data visualization tools—These are a class of tools that help to display the structured and unstructured data in people friendly and visual ways through graphs, plots, tables, charts, and statistical graphs, which enhance the understanding of the data trends and potential business implications. Some common uses are geographic information system maps, charting libraries to provide an array of visualization options, and mobile enabled visual representations.

Data mining tools—These tools come with inbuilt algorithms that can be used to analyze large data sets of enterprise data to discover unknown patterns in the data that help to understand business queries such as customer propensity, production affinity as well as the detection of anomalies.

Enterprise portal tools—These tools are help with the integration of information and people across enterprise boundaries. The key feature is that the enterprise portals have a secure unified access point through which employees, customers, and suppliers can access relevant information for their individual needs. Corporate portals or intranets however are limited to only employees or contractors of an enterprise.

Some of the key architecture principles around an information delivery and consumption layer would include the following:

- The information delivery and consumption solution should provide the organization a trusted, unified, and consistent view of cross-functional information and controls distribution of enterprise data assets across the enterprise in a standardized manner.

- The information delivery and consumption solution should provide the right data to the right audience at the right time.

- The information delivery and consumption solution should provide information to business partners and customers through a secure mechanism where only the required access is provided for their respective needs.

- In a big data enabled world in which enterprises dabble with both structured and unstructured data, the information delivery and consumption solution should be able to provide data visualization capabilities for all types of data irrespective of source.

- The information delivery and consumption solution should provide data to end users through a series of delivery channels. The user experience should be equivalent across the delivery channels.

- The information delivery and consumption solution should provide a security model that maps user data access to user job role. It also should factor in external data users such as customers, suppliers, and vendors.

The solution outline phase is when the tool evaluations are finished and the required tools are selected. Here I briefly cover the process of information delivery and consumption tool evaluation. There are a number of criteria based on which information delivery and consumption tools need to be evaluated. The evaluation criteria can be broadly classified into two areas namely,

- Functional capabilities

- Vendor capabilities—Country support and product support

See Table 9-2 for information delivery and consumption tool evaluation criteria.

Table 9-2. *Information Delivery and Consumption Tool Evaluation Criteria*

Evaluation Area	Evaluation Criteria
Functional: Business intelligence	Support for different report types (parametrized, tabular, crosstab, freeform, scheduled, ad hoc), report formatting capabilities (highlighting rows/columns, embedding logos, inserting page breaks),ease of use, online help, support for WYSIWYG ("what you see is what you get") printing for both web and client, report publishing in multiple formats, standard report templates, metadata layer to allow users to use business terms in row and column names, support for APIs and programming, integration with OLAP (online analytical processing) tools and portals, support for slicing and dicing, user interface support for rich visualization through 3D histograms, charts and graphs, report bursting, complex analysis and calculations, ability to save and reuse calculations and others.
Functional: Business activity monitoring	Support for mission critical business functions, ability to operate in fault tolerant mode, visually interactive dashboards with personalization features, delivery of alerts in multiple formats (SMS/e-mail), scripting tool to support actions in the context of the events or alerts, web links that provide access to detailed information about the events or alerts, support for historical reporting and analytics, knowledge capture features to record situations and actions, search functionality, ability to design custom dashboards, library of analytical functions (such as static and dynamic thresholds, absolute values, frequency, etc.),complex event processing, ability to integration with third-party rules engines.

(continued)

Table 9-2. (*continued*)

Evaluation Area	Evaluation Criteria
Functional: Planning and budgeting	Support for integration with strategy mapping tools to map the overall planning process, ability to create and preserve scenarios, ability to create multiple versions of budgets with different scenarios, ability to set target across multiple departments, ability to allocate targets across the entire hierarchy, ability to consolidate targets at all levels, ability to allocate targets based on rules, ability to allocate targets based on previous year targets, ability to set expense budgets across departments, ability to allocate expense budgets across the entire hierarchy, ability to consolidate expense budget at all levels, ability to allocate expense budget based on rules, accommodate zero based budgeting paradigm, ability to switch between top-down and bottom-up planning and budgeting processes, ability to create workflows, ability to write back targets and expense budgets to the accounting and transaction systems, baseline and version budgets, Excel-like front end for business users.
Functional: Scorecards and dashboards	Ability to integrate with strategy mapping tools to derive KPIs, ability to define and maintain KPIs, ability to define targets and ranges for KPIs and metrics, ability to visualize KPIs (using gauges, dials, gauges, thermometers), generate heat maps, ability to embed dashboards/scorecards in MS Office tools, ability to integrate with third party widget tools.
Functional: Data visualization	Ability to support diverse types of interactive graphical displays, analytics capabilities, easy report building, in-memory processing capabilities, ability to distribute reports/dashboards through mobile devices and portals.
Functional: Data mining	Support for a graphical user interface based intuitive user interface, integration with database management systems, ability to perform different types of analysis (association, clustering, classification, outlier analysis, regression, forecasting, etc.),ability to analyze text, data cleansing and transformation, API interfaces, ability to integrate with OLAP tools, integration with Excel spreadsheets, etc.
Functional: Portals	Support for languages in case of global user base, integration with database management systems, integration with OLAP/reporting tools, create, administer, and maintain forums, document management, integration with WAP (wireless application protocol) based services, search functionality, document indexing, provide secure access to documents and information based on user credentials, trigger event based alerts, integration with e-mail systems such as Lotus Notes and Microsoft Outlook.
Nonfunctional: All tools	Largest instance different licensing modes and costs
Vendors capabilities	Collaborating with hardware and software vendors
Vendor: Product support	Training support provided quality of product documentation support for new upgrades and patches; availability of migration path from older to newer product versions
Vendor: Country specific support	Vendor presence in the country of implementation; number of active clients and case studies; number of vendor consultants in the given country; number of consulting partners in the given country

Critical Success Factors in Information Delivery and Consumption

Despite numerous efforts and investments in information delivery and consumption initiatives there are numerous instances where such programs have failed to deliver and produce the desired benefits. Although embarking on an information delivery and consumption initiative with a maturity assessment to understand current capabilities and gaps is a good starting point, there is a need for a certain direction to sustain the momentum and realize business benefits. In this section, the critical success factors to implement an information delivery and consumption solution are covered.

As discussed in the section on building blocks, there is a sequence of activities performed when embarking on an information delivery and consumption program (see Figure 9-5).

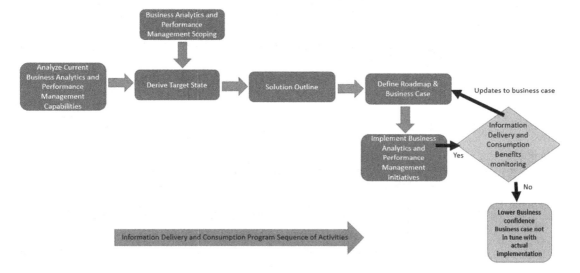

Figure 9-5. *The information delivery and consumption program sequence*

The sequence of activities include analyze current state and business analytics and performance management scoping ➤ define target state and current gaps ➤ solution outline (solution architecture and tools) ➤ define road map and business case (initial) ➤ implement information delivery and consumption initiatives that are defined based on the roadmap and monitor business benefits and feedback into the business case for a more realistic understanding of the business case. The key critical success factors are as follows:

> Build on the road map and business case—One of the key deliverables of the initial information delivery and consumption strategy is an implementation road map (with a list of business analytics and performance management initiatives to support the enterprise business vision) and a business case.

> Define performance management framework—It is vitally important that the performance management framework is defined as part of the information delivery and consumption assessment. The performance management framework could be a Balanced Scorecard methodology or Malcolm Balridge framework. Once the framework is defined the focus is to identify the key result areas in each of the Balanced Scorecard quadrants such as financial, customer, internal (operations), and learning and growth.

Design value driver trees and visualization strategy—It is important to design value driver trees that link the key performance areas for the business to the key performance indicators, providing a feedback mechanism for business performance. I provide a real-life value tree for illustration (see Figure 9-6).

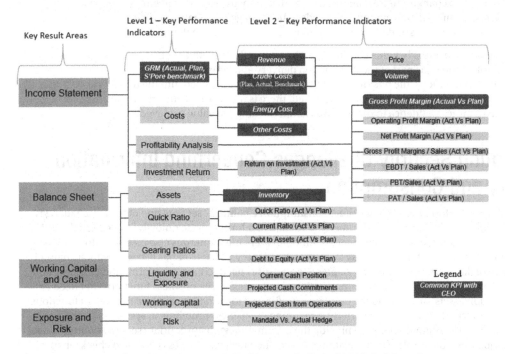

Figure 9-6. *Financial performance tree—Chief financial officer (CEO) dashboard*

For each of the business quadrants in which performance needs to be measured against the enterprise strategy and vision, the key result areas are identified. The key performance indicators are identified for each of the Key Result Areas (KRAs) as shown in Figure 9-6. The KPIs can be at various levels: top level KPIs, which are displayed on the dashboards, are known as level 1 KPIs and the next level of KPIs as level 2, and so forth. For instance for the income statement KRA, Gross Refining Margin (GRM) is a key KPI and is calculated as revenue derived from petroleum products coming out of a refinery minus total crude costs, which are the raw material costs.

It also is crucial to define the visualization strategy that helps to identify the scorecards and dashboards by function area (e.g., finance, operations, HR), classify the KPIs that define the scope of a scorecard or dashboard, define mock-ups for the key scorecards and dashboards that define the look and feel and seek stakeholder feedback, and perform a similar exercise for the key reports by functional area. Visualization strategy ensures that business stakeholders are engaged and there is an early agreement on KPIs that are crucial from a BI and performance management perspective. This also enhances the stakeholder buy-in for BI and performance management initiatives.

Build incrementally—Although there may be multiple business analytics and performance management initiatives defined as part of the information delivery and consumption implementation roadmap, it is important to understand that the initiative involves people, process, and technology; and an incremental build is always a more pragmatic approach to deliver incremental business value. For instance, scorecards and dashboards can be built first, and then planning and budgeting applications can be implemented.

Information delivery and consumption as part of information governance—The need for information governance is crucial to make an information delivery and consumption initiative succeed. KPI definitions, calculations, and data sourced needs to be governed as well as data consumption processes that manage exchange of information assets both within and outside the enterprise. The exchange of information also needs to be managed through enterprise governance teams and ownership for information assets to be defined.

Monitor business benefits continually—As each of the initiatives get implemented (as part of the information delivery and consumption implementation), the business benefits achieved need to be fed into the initial business case. When that is accomplished, there is real-world feedback, which gives more confidence to the business about the tangible value achieved.

Information Security Challenges Concerning Information Delivery and Consumption

As discussed in the "Information Delivery and Consumption Channels" section, one of the key challenges transnational enterprises face today is the ability to exchange information seamlessly with external entities such as suppliers, vendors, and even customers. Although there are numerous channels for information consumption, it is crucial to have a security framework to identify which user has access to which channel and the nature of data sets. As defined in Chapter 6, enterprises need to have an information consumption security process model that helps classify information assets, analyze the business functions that need to access information assets, understand the employee profile and job function, analyze the system functions that access the information assets, and perform a mapping of business functions to systems accessing key information assets. The ultimate goal is to provide the information with appropriate security controls based on a person's job function or role. The mitigation steps include reviewing access control to check whether users have the right access control over information assets, audit systems, information channels that provide access, develop training, and enablement and contingency plans.

Tools for Information Delivery and Consumption

After reviewing the information delivery and consumption building blocks and critical success factors, I cover some of the tools in this domain. Some of the key tools include BI (includes scorecards and dashboards), business activity monitoring, planning and budgeting, data visualization, data mining, and change enterprise portal tools. The vendors covered are the ones that have capabilities in all the areas mentioned. Some of the market's leading tools in each of these categories are covered in Table 9-3.

Table 9-3. *Tools for Information Delivery and Consumption Solutions*

Information Quality	Oracle	IBM	SAP
Business intelligence	OBIEE	Cognos	SAP BusinessObjects
Business activity monitoring	Oracle BAM	IBM Business Monitor	SAP Solution Manager
Planning, budgeting, and forecasting	Hyperion Planning	Cognos Planning, Cognos TM1	SAP Business Planning and Consolidation
Data visualization	OBIEE (Oracle ADV)	Infosphere Data Explorer	SAP Lumira
Data mining	Oracle DataMiner	SPSS	SAP BW Data Mining Analytics
Enterprise portals	Oracle WebLogic Portal	WebSphere Portal Server	SAP Enterprise Portal

CHAPTER 10

■ ■ ■

Pillar No. 7: Metadata Management

Now that you understand the major components of an enterprise information management (EIM) solution, in this chapter I explain metadata management. Metadata management helps enterprises make sense of the information that is contained in their corporate repositories. *Metadata* is often referred to as "data about data" and in this chapter I look at the crucial role of metadata in EIM. As with information governance, metadata management cuts across the spectrum of EIM from information sourcing to information delivery and consumption. Metadata enhances the business and technical value of information stored in enterprise repositories such as operational data stores (ODSs), information warehouses, and data marts.

The primary purpose of metadata management is to provide context and semantics for enterprise data as it moves through the information supply chain from information source to information delivery and consumption. Metadata management is the discipline that deals with managing the semantics and context of the data from a business, technical, and operational perspective. In this chapter the following topics are covered:

- *Metadata management definition*

- *The key drivers for metadata management*

- *Building blocks for metadata management*

- *Critical success factors for metadata management*

- *Tools for metadata management*

The next several sections talk about each of these key considerations in more detail.

■ **Note**   The chapter goal is to explain metadata management, the key drivers for metadata management, and how organizations can build foundations for effective metadata capability. Also discussed are the critical success factors to consider when embarking on metadata management initiatives.

Although I introduced the concept of metadata management, it is also important to consider the supporting disciplines of information governance that enhance the significance and value of the metadata through timely integration and the robust processes needed for metadata being kept up-to-date with information integration, information warehousing, and information delivery and consumption changes.

Metadata Management Definition

Although there are numerous definitions of metadata management, a practical one would be—metadata management is a discipline that deals with the semantics and context of data as it is generated from a source system through integration to information warehouses and data marts, delivery, and consumption, through channels to information consumers and then on to disposition and retirement. At each stage of the information supply chain, metadata or semantics and context about enterprise data is generated and captured through a metadata management solution. The metadata generated at each stage can be business, technical, operational, or a combination of all types. Metadata is data about data that helps answers questions such as where is the data used, what is the business definition of a data element, when was the data element last updated, how is the data element related to other data elements, and so forth.

Some of the more conventional definitions include the classic Gartner definition of metadata management—metadata is information that describes various facets of an information asset to improve its usability throughout the information lifecycle. It is metadata that helps turn information into an asset.

Closely related to metadata management is the discipline of information governance. Information governance is an enterprise-wide initiative to control and audit data management processes as well as to ensure ownership of information assets. Metadata management is one of the key pillars for enabling information governance. To ensure that the enterprise data assets are trustworthy, the data assets should be catalogued and identified, and that requires governance for compliance, security, and privacy or exchange with business partners. Effective metadata management ensures a consistent definition of enterprise data assets that in turn promotes information sharing across units, which enhances business integration and helps provide a 360 degree view of key business entities such as customers and products. Sharing of data also benefits the customer experience as feedback from customer contact centers can be fed into supply chain efficiencies and order management processes. Figure 10-1 shows the types of metadata generated at each stage of the information lifecycle.

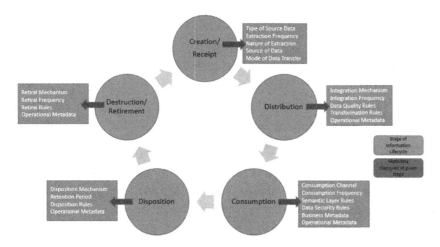

Figure 10-1. *Metadata captured at each stage of the information lifecycle*

As seen in Figure 10-1, metadata is captured at each stage of the information lifecycle. There are three types of metadata that add context and semantics about data as it moves through the information lifecycle, they are—technical metadata, business metadata, and operational metadata. Business metadata deals with the definitions of measures and serves as a business glossary where business users and analysts can look up the measures. This provides an enterprise-wide glossary of business definitions that often results in a better understanding of enterprise data assets. Technical metadata captures technical information about the source system such as extraction mechanism, data transfer mechanism, transformation rules, data quality rules, data model specifications, and so forth. Technical metadata provides insights about the technical design of a solution and how the data moves across the information supply chain. Operational metadata provides insights and statistics about the nature of queries asked by business users; report usage and adoption trends; run time of extract, load, and transform (ETL) jobs, and job statistics about the number of records processed; number of records rejected; and so forth. Operation metadata is captured right through the information lifecycle and provides insights into the operations concerning the data processing and consumption throughout the information lifecycle.

Key Drivers for Metadata Management

The key drivers for metadata management from an enterprise standpoint are as follows (see Figure 10-2):

- Enhance business productivity
- Improve change management
- Cost optimization
- Enhance business collaboration
- Enhance IT productivity
- Reduce compliance risks

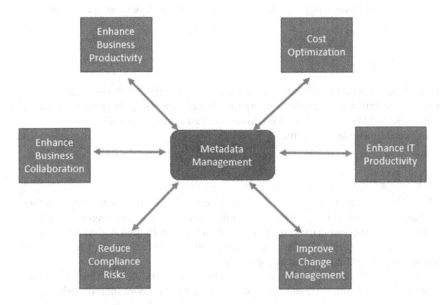

Figure 10-2. *Business drivers for metadata management*

Enhance Business Productivity

One of the key drivers of metadata management is the measure of trust provided to business users about the underlying enterprise data assets through data lineage and impact analysis. Businesses can learn the accuracy, comprehensiveness, and currency of data in the enterprise system of record. This in turn can enhance business productivity.

Improved Change Management

Metadata management provides enterprise-wide visibility into data definitions, data lineage, and impact analysis of changes brought about by changing business and system requirements. Metadata management provides the ability to perform accurate and timely impact analysis to understand the impact of changes. This ensures that there is improved ability in the handling of change management.

Cost Optimization

Metadata management provides enterprises with an end-to-end view of enterprise data assets and their definitions as well as end-to-end data lineage across the information supply chain. This enhances productivity in the enterprise in the following areas:

- The ability to exchange data across business units leading to better decision making and faster turnaround time, which enhances customer service and better execution of operations.

- The agility to handle changes due to the better understanding of the impact on the enterprise.

- Business and IT productivity through metadata management tools that provide insights concerning data assets in the inventory, and how the data assets can be utilized for business decision making.

All of these translate into cost-optimization opportunities from a human resources and operations standpoint.

Enhance Business Collaboration

Metadata management ensures that consistent definitions of data elements and their end-to-end lineage results in greater data sharing across business processes and business units. This results in greater business collaborating, which in turn results in the optimization of business processes throughout customer relationship management and operations management.

Reduce Compliance Risks

One of the key business drivers for metadata management is to reduce compliance risks that enterprises face in numerous industries such as banking, insurance, health care, and so forth. Metadata management as a discipline provides an enterprise with data assets that are consistently defined, with their end-to-end data lineage and transformation rules clearly defined. This provides data assets with traceability and an audit trail and provides a stronger, robust framework for compliance reporting.

As metadata management clearly catalogs and describes enterprise data assets, the semantics reduces the risk of end users and business analysts selecting the wrong data or filters for a report, thereby reducing the risk.

Enhance IT Productivity

In the absence of up-to-date design documentation, metadata provides vital information about the underlying systems and data structures thereby enhancing the IT productivity. Source system changes and their impact on existing ETL jobs, reports, or assessment of the impact of changes due to alterations in measure definitions can be easily performed using a metadata management tool with a unified metadata repository. This saves a considerable amount of analysis and data collation time, enhancing the IT productivity.

According to a TDWI survey conducted in 2010, the leading enterprise applications for metadata management included information warehousing, information integration, information quality, information governance, master data management, and customer data integration. The wide range of information management applications that leverage metadata management highlights its crucial contribution to enterprises as a supporting information management discipline.

Metadata adds to the value of enterprise data and helps answer both business and technical queries about the data. Some of the common queries that can be addressed by a robust metadata solution include the following: how old is the underlying data, when was the data last refreshed, which data elements are customer specific and have privacy concerns, and which emission norms need to be catered for a specific country, and so forth.

Building Blocks and Enablers for Metadata Management

With an understanding of what is metadata management, I move on to the building blocks of metadata management within an enterprise. Enterprises should build a metadata management business case and roadmap based on an assessment of metadata needs in the enterprise, and target future capabilities based on the enterprise vision. The six phases of a metadata management strategy are engagement initiation, metadata requirements scoping, assess current metadata capabilities, define target state, solution outline, and define metadata road map and business case. Figure 10-3 shows the different phases of a metadata management strategy. Table 10-1 then defines the key activities and deliverables for each of the metadata management strategy phases.

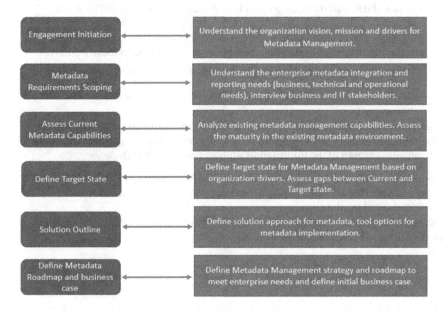

Figure 10-3. *Phases of a metadata management strategy*

Table 10-1. *Activities and Deliverables for Metadata Management Strategy Phases*

Engagement Phase	Activities and Deliverables
Engagement initiation	Engagement kickoff meeting with stakeholders. Define work plan and workshop schedules. Define the metadata maturity assessment parameters, solution templates, and project governance committee.
	Deliverables: Project plan, interview/workshop schedule, templates
Metadata requirements scoping	Interview business and IT stakeholders to understand organization vision around metadata management. Derive the business, technical, and operational metadata requirements.
	Deliverables: Metadata requirements scoping document
Assess current metadata management capabilities	Analyze existing processes, skills, and technologies related to metadata management. Assess the maturity in the existing environment with respect to defined metadata assessment parameters.
	Deliverables: Current state metadata management assessment report
Define target state	Conduct metadata capabilities workshops to derive enterprise metadata vision. Define target state architecture to support the required capabilities. Assess gaps to be addressed in the current state to achieve the target state.
	Deliverables: Metadata capabilities and target state architecture document
Solution outline	Derive solution options for the metadata management solution. Perform tool evaluations and proof of concepts.
	Deliverables: Technology options with recommendations
Define metadata road map and business case	Define the metadata management initiatives with business and technical priorities. Define the implementation road map based on feedback from business and IT stakeholders. Define initial business case for first set of metadata initiatives and define the next steps to implement these initiatives.
	Deliverables: Metadata management implementation roadmap and business case

With an understanding of how to get started with metadata management at an enterprise level, I start with the building blocks of a metadata management solution. Figure 10-4 covers the key components of the solution, each contribute to metadata as part of the information supply chain. These include,

> Data integration engines—Ingest and extract data from numerous source systems (internal and external) and load the data into information warehouses, data marts, and ODSs. All of the data integration metadata (technical and operational) is captured into the unified metadata repository, which is the single unified source of end-to-end metadata in the information supply chain.

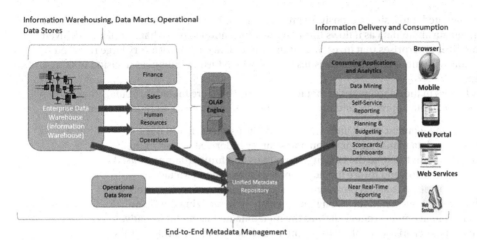

Figure 10-4. *Building blocks of a metadata management layer*

Information warehouse, data marts, and operational data stores—The metadata associated with enterprise decision management repositories, such as information warehouse, data marts, and ODS, provides valuable insights into enterprise data assets. Metadata is a combination of business, technical, and operational metadata. This provides a data lineage into how the data elements are derived from their respective sources, the transformation rules, the underlying data models, measure definitions, dimension hierarchies, and operational details about the frequency of refresh and data quality trends. All of this metadata is integrated into the unified metadata repository through a set of metadata bridges.

OLAP/semantic layer—The OLAP/semantic layer is often designed to have a layer in which the multiple enterprise repositories are connected, for example, the information warehouse, data marts, and facts and measures are exposed to business users in business friendly terms, hiding the data model naming conventions and technical complexities. The OLAP layer provides metadata about the nature of dimension hierarchies, custom hierarchies, aggregation modes and methods, drill down and drill access paths, and so forth. Business and technical metadata can be integrated using metadata bridges into the unified metadata repository.

Information delivery and consumption layer—The information delivery and consumption layer contains crucial metadata concerning the visualization of data, key measures, and key performance indicators (KPIs) displayed on scorecards and dashboards, the frequency of the refresh of the data, where the measures are leading or the lag indicators, underlying security metadata about who has access to these measures and KPIs, calculation logic of the measures and KPIs, report bursting schedules, mechanism of reporting delivery by user, user personalization features, and so forth. Business, technical, and operational metadata related to information delivery and consumption can be integrated into the unified metadata repository.

All metadata integrated into a single repository produces an end-to-end store of business, technical, and operational metadata about data as it flows through the information supply chain until it is archived and destroyed. This gives enterprises that invest in a unified metadata repository a very large reservoir of metadata that provides semantics and characteristics of data, which further enhances decision making and quality of outcomes.

The three key factors in the enablement of metadata management are metadata management vision and strategy, unified metadata approach, and metadata management solution architecture and tools.

Metadata management vision and strategy—As discussed earlier in the chapter, metadata management vision and strategy is derived as part of the initial assessment to gauge the current capabilities and future roadmap based on enterprise vision and future metadata management needs. Metadata management vision is based on an enterprise's mission or vision statement and understanding how metadata management can help in realizing the vision. Metadata management strategy defines the metadata management initiatives that help an enterprise to achieve its vision. In a close link to the strategy is the metadata management road map and business case that help to determine the sequence of initiatives and the business benefits the metadata the initiatives provide. The business case is defined upfront and needs to be monitored throughout the lifecycle of the metadata implementation to provide a means of continuous benefits.

Unified metadata approach—The unified metadata approach is the mechanism through which enterprises define an unified metadata repository for all types of metadata that are part of the data generated from the various business transactions and processes. Enterprises need to discover how to integrate the diverse metadata assets into a single repository to provide end-to-end data lineage capabilities, insight into the data, and single and consistent definitions of key business data elements. The potential for large numbers of enterprise applications, such as master data management, customer data integration, information warehousing, and so forth, to tap into this unified metadata repository should be enough incentive to invest in an unified metadata approach.

Metadata management solution architecture and tools—A key driver in any metadata management solution is to define the solution architecture and it's supporting tools and technologies. There are two key aspects of solution architecture that need to be defined: 1) identify the solution components needed to build metadata management capabilities in a given enterprise, and 2) identify the tools and technologies needed to build the metadata management solution. The key solution components that go into building a metadata management solution are as follows: 1) unified metadata repository; 2) metadata integration tools for the integration of business, technical, and operational metadata; 3) business glossary-like tools; and 4) impact analysis tools for performing data lineage analysis and impact analysis of changes.

Unified metadata repository—The metadata solution must have a central repository for storing all types of metadata—business, technical, and operational. The unified metadata repository must have a metamodel for storing the diverse types of metadata.

Metadata integration tools—For the unified metadata repository to store all types of metadata (business, technical, and operational) there needs to be metadata integration tools/bridges through which the metadata can be integrated, such as information integration tools for data integration metadata, data modelling tools for data model related metadata, BI tools for information delivery and consumption metadata, and OLAP tools for semantic layer metadata. The metadata integration tools also need APIs for the integration of metadata from third party tools that may have a different metamodel. Although import bridges can provide partial integration, a custom code may need to be written, making the process more complex and difficult to maintain.

Business glossary tools—Business glossary-like tools serve as a glossary for common business terms and measure definitions that promote the usage of data elements across business units and promote data collaboration in enterprises. Business glossary-like tools enable business analysts, data analysts, and data stewards to work together to create, manage, and share a common understanding of business terms. Business glossary-like tools enable business users to be up-to-date with the data elements and business terms available in the system of record and other decision management repositories. This enables larger usage and ownership of data elements in the enterprise.

Impact analysis tools—Reporting and impact analysis tools in the metadata solution enable business analysts and data analysts to analyze data lineage as well as the traceability of data elements as they move across the information supply chain. Most metadata management tools come with prebuilt reports for doing data lineage analysis and the study of interrelationships of data elements.

Some of the key architecture principles around metadata management would include the following:

- The metadata management solution should provide the organization with a trusted, unified, and consistent view of metadata integrated from enterprise applications.

- The metadata management solution would provide end-to-end impact analysis and data lineage capabilities from data sourcing to information delivery and consumption.

- The unified metadata repository would serve as the catalog for the diverse types of metadata. Metadata repositories can be used in an enterprise context or in a wider context such as a core standard or ontology, extensions of a standard, usage within a domain such as access to schemas of interest to a specific domain such as education, supply chain, and so forth.

- In a big data enabled world in which enterprises dabble with both structured and unstructured data, the metadata management solution needs to have metadata associated with both structured and unstructured data.

- The metadata management solution would provide business users in an enterprise (private or public) with a glossary of relevant business terms that serves as a reference for key business definitions relevant to the business processes.

The solution outline phase is when the metadata tool evaluations are finished and the required tools are selected. Here I briefly cover the process of metadata management tool evaluation. There are a number of criteria based on which metadata management tools need to be evaluated. The evaluation criteria can be broadly classified into two areas namely,

- Functional capabilities

- Vendor capabilities—Country support and product support

See Table 10-2 for metadata management tool evaluation criteria.

Table 10-2. *Metadata Management Tool Rvaluation Criteria*

Evaluation Area	Evaluation Criteria
Functional: Unified metadata repository	Support for centralized/federated/hybrid repository for storage of business, technical, and operational metadata.
Functional: Metadata integration tools	Metamodel that supports common warehouse metamodel (CWM). CWM specifies interfaces that can be used for the interchange of metadata between the information warehouse, BI, and data model tools. CMW is based on three standards—UML (unified modeling language), MOF (meta-object facility), and XML (XML metadata interchange).
Functional: Business glossary tools	Support for business glossary capabilities that allow common business terms to be defined and stored in a glossary/dictionary. Support for search and navigation features in the business glossary with thin client/web access for wider usage/adoption. Ability to link business terms and categories to underlying logical data models and tables.
Functional: Impact analysis and reporting	Support for end to end data lineage and impact analysis capabilities. Inbuilt reports for impact analysis capabilities of changes due to source system changes or measure definition changes.
Vendors capabilities	Largest instance Different licensing modes and costs Collaborating with hardware and software vendors
Vendor: Product support	Training support provided Quality of product documentation Support for new upgrades and patches Availability of migration path from older to newer product versions
Vendor: Country specific support	Vendor presence in the country of implementation Number of active clients and case studies Number of vendor consultants in the given country Number of consulting partners in the given country

Some of the common industry standards for metadata exchange include the common warehouse metamodel (CWM) standard. The purpose of CWM is to enable the interchange of CWM between tools and metadata repositories in heterogeneous environments. CWM is the specification of syntax and semantics that information warehousing and BI tools can leverage to interchange shared metadata. Meta-object facility (MOF) is the modeling language, while UML is the modeling notation and the base metamodel, and the XMI standard allows metadata to be interchanged based on XML as the interchange mechanism.

Critical Success Factors in Metadata Management

Although embarking on a metadata management, enterprise-wide program with a maturity assessment to understand current capabilities and gaps is a good starting point, there is a need for a certain direction to sustain the momentum and to realize the business and technical benefits. In this section, the critical success factors in implementing a metadata management solution are covered.

As discussed in the section on building blocks, there is a sequence of activities performed while embarking on a metadata management program (see Figure 10-5).

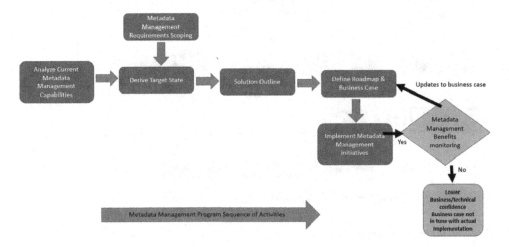

Figure 10-5. *Metadata management program sequence*

The sequence of activities include analyze current state and metadata management requirements scoping ➤ define target state and current gaps ➤ solution outline (metadata solution architecture and tools) ➤ define road map and business case (initial) ➤implement metadata management initiatives that are defined based on the roadmap and monitor business/technical benefits and feedback into the business case for a more realistic understanding of the business case. The key critical success factors are as follows:

> Build on road map and business case—One of the key deliverables of the initial metadata management strategies is the implementation road map (with a list of metadata management initiatives to support the enterprise business vision) and a business case.

> Metadata process design—One of the key gaps in enterprises is the lack of processes that capture metadata needs from the start of an information management program to the archival and retirement of associated data elements. Metadata processes need to be designed to capture the different types of metadata (see Figure 10-6) at different stages of an information management program. Metadata processes also need to look at metadata reporting requirements and ensure that the unified metadata repository is designed keeping such requirements in mind. In any information management initiative, business requirements are first derived and submitted to the enterprise information management center of excellence (EIM CoE); a construct used in many large enterprises today and looked at in detail in Chapter 12). The EIM CoE program office analyzes the business requirements and passes on the request

to the change control team of the metadata operations team. The changes, if approved, are then translated into metadata requirements, which are business metadata needs that in turn are translated to technical metadata and operational metadata requirements. Once these changes are implemented the metadata are integrated into the unified metadata repository to be monitored and controlled for future changes and requirements.

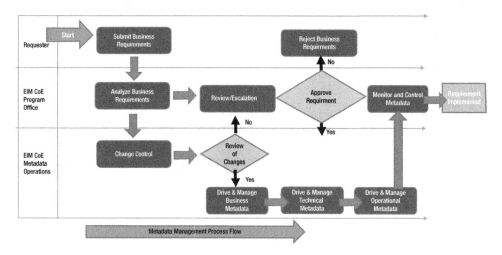

Figure 10-6. *Metadata management process flow*

Metadata management as part of master information management—Often metadata initiatives are kick-started due to an enterprise master information management initiative. This results in the key master data entities and their hierarchies captured as business metadata and the master data model and the technical design captured as technical metadata.

Metadata management as part of information warehouse—Often metadata initiatives are driven from information warehouse implementations. The business metadata related to the KPIs and measures are captured into a business data dictionary or glossary. The data model and ETL design specifications are captured as technical metadata and the ETL and reports operations data are captured as the operational metadata post go live.

Build incrementally—Although there may be multiple metadata initiatives that need to be built, it is practical to design and build a business glossary first as it ensures that common data definitions are finalized. A business glossary helps build data collaboration across business units. The business glossary implementation also ensures that there is a greater ownership of data assets and a better understanding of the business measures already implemented in the information warehouse, data marts, and other decision management platforms. Once the business metadata is integrated, the technical and operational metadata can be integrated into the unified metadata repository in subsequent initiatives.

Metadata management as part of information governance—Metadata initiatives are sometimes kicked-off as part of information governance programs to ensure that there is governance and ownership of enterprise data assets and consistency in data definitions. Data is cataloged using metadata management processes and provides end-to-end data lineage and analysis capabilities.

Monitor business benefits continually—As part of metadata implementation as each of the initiatives become implemented, the business/technical benefits achieved need to be fed into the initial business case for there to be real-world feedback concerning the numbers realized, which gives more confidence to the business and IT about the tangible value achieved.

Tools for Metadata Management

After reviewing the metadata management building blocks and critical success factors, I now discuss some of the tools in this domain with some of the key tools for unified metadata repository and management, business glossary, metadata integration, and impact analysis. The vendors covered are the ones that have capabilities in all the areas mentioned. Some of the market leading tools in each of these categories are covered in Table 10-3.

Table 10-3. *Tools for Metadata Management Solutions*

Information Quality	Informatica	IBM	SAP
Unified metadata repository	Informatica Metadata Repository[a]	Infosphere Metadata Repository[a]	SAP BI Metadata Repository Manager
Business glossary	Informatica Business Glossary	Infosphere Business Glossary	SAP BusinessObjects Information Steward
Metadata integration	Informatica Metadata Manager	Infosphere Metadata Asset Manager	SAP BusinessObjects Information Steward
Impact analysis	Infosphere Metadata Manager	Infosphere Metadata Workbench	SAP BusinessObjects Information Steward

[a]*Part of metadata solution not separate products.*

■ ■ ■

CHAPTER 11

Pillar No. 8: Big Data Components

We live in a world that is increasingly connected, instrumented, and integrated. Huge volumes of data are generated worldwide that are both structured and unstructured. Estimates based on numerous studies show that unstructured data generated from numerous sources, such as social media, sensors, machines, RFID tags, weblogs, and so forth, constitute 80% of the data generated worldwide. To put things in perspective, 30 billion pieces of content were added to Facebook by over 600 million users in October 2015 and 2.9 million e-mails are sent worldwide every second. Around 30 billion searches are created on Twitter in a month. Traditionally enterprise information management (EIM) has looked at only the structured data, but with the increasing emphasis on analyzing new data sources and data types to provide unique insights concerning product feedback and customer service, more and more enterprises are looking at ways to analyze such data sets. This brings us to the world of "big data" and how big data solution components are now a large part of the EIM landscape of enterprises in transformation.

Big data has mainly three perspectives—huge volumes, diverse data types, and speed of processing. This has an effect on the EIM solution architecture that needs to be extended to cater to these perspectives of volume, variety, and velocity. There are two other dimensions of big data: 1) at rest (as in information warehouses, data marts) and 2) in motion (streaming data from different sources that are analyzed in motion and not stored in a repository). In this chapter the following topics are covered:

- *Big data definition*
- *The key drivers for big data solutions*
- *Building blocks for big data solutions*
- *Critical success factors for big data*
- *Tools for big data*

The following sections discuss each of these key considerations in more detail.

■ **Note** The chapter goal is to explain big data, its key drivers, how organizations can build foundations for effective big data capability, and the critical success factors to consider when embarking on big data initiatives.

Although I introduced the concept of big data, it also is important to understand that the problems of dealing with big data have existed in many industries before the advent of big data technologies. For instance the oil and gas industry exploration companies have been dealing with a deluge of data from oil wells both during exploration as well as production. The data from oil rigs include sensor data, seismic 3D and 2D images, and well log data (both unstructured to semistructured), and involve a huge volume of data and require rapid processing to ensure that the rig operations and drilling teams get real-time insight from the analysis. Stock exchanges deal with real-time stock indexes that require processing large data volumes in real time. With the advent of Hadoop and MapReduce and later generations of big data technologies, enterprises are now enhancing their EIM capabilities with big data processing and analytical capabilities.

Big Data Definition

It is often unclear to determine when an enterprise moves from a traditional information warehouse to a big data realm as the definition of big data varies from organization to organization as well as across industries. Although there are numerous definitions of big data, a practical one would be that big data is a discipline that deals with the processing, storage, and analysis of heterogeneous (structured/semistructured/unstructured) large data sets that cannot be handled by traditional information management technologies that have been used to process structured data. Gartner defined big data based on the three Vs: volume, velocity, and variety.

- *Volume*—Enormous volumes of data are generated today through the Internet and from sensors in numerous mechanical devices from cars to oil rigs.

- *Velocity*—Streaming data at high speeds from sensors, RFID tags, and smart meters derive the need for processing volumes of data in near real time.

- *Variety*—Today data comes in diverse forms such as structured data from enterprise applications, semistructured data from weblogs, unstructured data from text documents, e-mails, social media, and so forth.

In addition there are other dimensions around the frequency of data that can change with events or seasonality. Big data can be considered to be a combination of these characteristics that creates an opportunity for enterprises to gain a competitive edge in today's digitized landscape.

Big data as discussed can be from diverse sources. Figure 11-1 illustrates the common data types in the big data ecosystem.

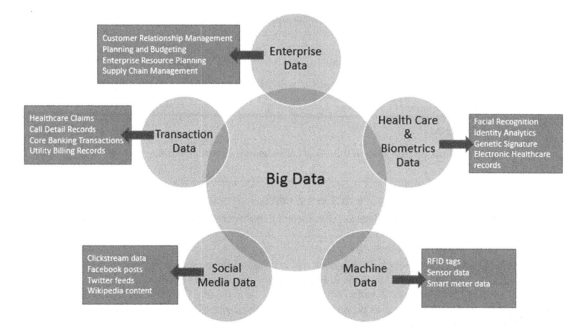

Figure 11-1. Common data types of big data

The common data types in the big data landscape include the following:

- *Enterprise data*—data from internal sources such as enterprise applications as customer relationship management (CRM) systems, enterprise resource planning (ERP), supply chain management systems, planning and budgeting applications, and so forth. These can be categorized as structured data.

- *Transaction data*—transaction data related to a specific industry such as health care claims, insurance claims, call detail records (CDR) in telecom, core banking transactions, utility billing records, and so forth. These can be both structured as well as semistructured in case of CDR.

- *Social media data*—social media data includes Facebook posts, Twitter feeds, online content, and clickstream data from web pages. These can be semistructured as in case of clickstream data and unstructured as in case of Tweets or Facebook posts.

- *Machine data*—Machine data is data generated from one machine to another and includes sensor data, RFID tags data, data from smart meters, and so forth. Machine data is usually unstructured.

- *Health care and biometrics data*—These include electronic health records (EHR) data, facial recognition, and identity data. EHR data is mainly unstructured data, including images from X-rays, free form text from physician notes, and so forth. Biometric data such as facial recognition is unstructured whereas identity data can be both structured as well as unstructured.

A suitable way to determine if an enterprise is big data ready is when its information landscape displays the following characteristics:

- Large data volumes from operations from a wide variety of sources such as enterprise applications, social media, machine data, weblogs data, weather data, and so forth

- More diverse data types (structured/unstructured/semistructured)

- Data that is retained for longer periods of time for different needs (government regulations, compliance needs). This results in enterprises looking for cheaper storage solutions leading to adoption of Hadoop as a storage option.

- Data used by a wide variety of applications from customer retention, loyalty and attrition analysis, impact of weather on sales, and so forth. In addition a combination of structured and unstructured data creates new analysis types such as integrating customer data in CRM systems with call center logs gives insights about voice of customer and customer satisfaction.

- Increasing pressures around time to market and faster decision making. This calls for new technologies that can ingest diverse data faster and produce insights and enhance decision making. In addition decision making involves linking of structured and unstructured data and the insights derived serve as inputs into product design and customer relationship management.

There are many real-world examples that exemplify how enterprises are now immersed in the big data realm. Utility companies need to forecast energy production to meet potential demand. To meet this need, wind power companies need to analyze weather data on a frequent basis. Wind power companies are able to analyze petabytes of weather data to determine placement of location site in minutes instead of weeks (as in the past; using traditional information management methods). Logistics companies such as UPS are using big data analytics to track packages and trucks. The data comes from sensors on the trucks, which helps with better route planning and in turn saves driver time and fuel consumption, resulting in cost savings. With an explosion of data with Twitter and Facebook processing terabytes of data on a daily basis, over 200 million smart meters the age of big data has arrived. Enterprises need to find the correct use cases to leverage this data as a natural asset for actionable insights that creates new opportunities for revenue generation, cost optimization, and quicker decision making.

Key Drivers for Big Data Solutions

As is evident from the previous section, big data is a transformational solution that can change the way an enterprise operates. The key drivers for big data solutions from an enterprise standpoint are as follows (see Figure 11-2):

- Data monetization opportunities

- New product innovations

- Deeper customer insights

- Operational process efficiencies

- Fraud detection and reduction of risk

- Cost optimization

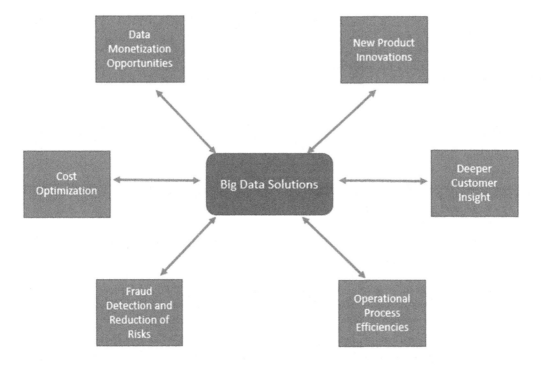

Figure 11-2. *Business drivers for big data solutions*

Data Monetization Opportunities

One of the key drivers of big data solutions is the ability to create new data monetization opportunities and new business models with revenue generating potential. For example by analyzing weather data, retailers can provide customers with in store promotions that are not impacted by weather events. Geospatial data analyzed with store footfalls and sales can help retailers fine tune their promotions. Reading from smart meters in near real time can provide utility companies with trends of energy consumption and provide energy discounts to customers during off-peak load hours. Analyzing CDR data can help telecom companies to fine tune their service levels by analyzing dropped calls and link them to customer attrition.

New Product Innovations

Enterprises with a big data solution have large data assets to analyze the social media trends about product usage, brand affinity, and customer acceptance for products and services. Analyzing this data with transaction data and order fulfillment data provides the enterprise with a picture about product usage and customer feedback and provides insights about new product innovations that would meet customer demand or create a new market.

Deeper Customer Insights

Big data solutions provide deeper insights about customer buying trends and attrition as they have the ability to link structured and unstructured data from diverse data sources from CRM systems, ERP systems, supply chain management (SCM) data to social media, and segmentation data. The deeper insights are attained through promotion analysis, clickstream analysis, customer profitability and life time value analysis, churn prediction, and loyalty program analysis. The huge data sets that can be analyzed in a relatively short time makes decision making faster and creates new opportunities for the business to engage effectively with the customer.

Operational Process Efficiencies

Big data solutions enable events to be tracked in near real time in operational systems creating opportunities to optimize processes as well as take corrective actions when certain events/alerts are triggered. The examples for operational process efficiencies are cross-industry including: patient monitoring in ICUs; optimization of transportation networks such as roads, railroads, and shipping routes; inventory optimization by linking sales with inventory and reordering processes; monitoring drill rigs in the oil and gas industry to manage drill operations; environmental analysis such as analyzing weather patterns that help wind farm companies, and so forth.

Fraud Detection and Reduction of Risk

One of the key business drivers for big data solutions is to proactively manage fraud detection by linking multiple data sets (structured and unstructured) and also reduce compliance risks through regulatory reporting. Big data solutions enable the integration of historical data with fraud modeling to enable the detection of patterns and also integrate identity data with surveillance, which enables robust fraud detection capabilities.

Cost Optimization

Big data solutions also create opportunities to cut costs for IT as Hadoop can be used to store both structured and unstructured data for long periods of time with no performance degradations in terms of data ingest or querying. This ensures that the storage costs diminish with no performance implications. In additional big data solutions enable integration of new data types that would be more complex and costly to integrate in traditional information integration methods.

Another increasingly key driver for big data solutions is the "Internet of things." Internet of things or IOT refers to a network of machines that have sensors and are interconnected enabling them to collect and exchange data. This enables devices to be sensed and controlled remotely resulting in process efficiencies and lower costs. As the IOT deals with huge volumes of data collected and exchanged over the IOT network it creates huge opportunities for big data solutions to make a tangible difference with its data ingest, analytics, and visualization capabilities.

Building Blocks and Enablers for Big Data Solutions

Now with a basic understand of big data, I move to the building blocks of big data solutions in an enterprise. Enterprises should build a business case and roadmap based on the assessment of potential big data use cases and needs in the enterprise and the targeted future capabilities based on the enterprise vision. The seven phases of a big data strategy are as follows:

- Engagement initiation

- Deriving big data use cases and requirements

- Assess current big data capabilities

- Define target state

- Big data solution outline

- Define big data roadmap and business case

- Big data pilot and apply learnings and define next steps

Figure 11-3 shows the different phases of a big data strategy.

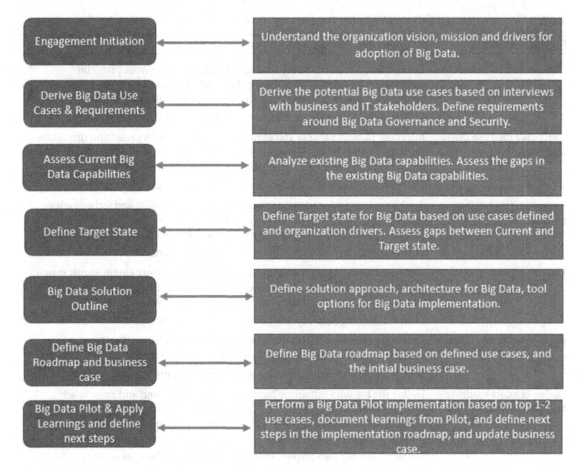

Figure 11-3. *Phases of a big data strategy*

Table 11-1 defines the key activities and deliverables for each of the big data strategy phases.

Table 11-1. *Activities and Deliverables for Big Data Strategy Phases*

Engagement Phase	Activities and Deliverables
Engagement initiation	Engagement kickoff meeting with stakeholders. Define work plan and workshop schedules. Define the big data maturity assessment parameters, solution templates, and project governance committee.
	Deliverables: Project plan, interview/workshop schedule, templates
Derive big data use cases and requirements	Interview business and IT stakeholders to understand organization drivers for big data. Derive the business and technical use cases around big data. Define big data governance and security requirements in line with the use cases.
	Deliverables: Big data use cases and requirements scoping document
Assess current big data capabilities	Analyze existing information management processes, skills and technologies to assess the maturity in the existing environment with respect to big data parameters. Assess gaps/barriers in the current landscape for big data adoption.
	Deliverables: Current state big data assessment report
Define target state	Define target state based on big data use cases and vision. Define target state architecture to support the required big data capabilities. Assess gaps to be addressed in the current state to achieve the target state.
	Deliverables: Big data capabilities and target state document
Big data solution outline	Derive solution options for the enterprise big data solution. Perform tool evaluations and define solution architecture for big data.
	Deliverables: Technology options and solution architecture document
Define big data roadmap and business case	Define the big data initiatives based on use cases defined and associated business and technical priorities. Define the implementation roadmap based on feedback from business and IT stakeholders. Define initial business case for first set of big data initiatives.
	Deliverables: Big data implementation roadmap and business case
Big data pilot, apply learnings, and define next steps	Implement a big data pilot based on the top one to two use cases defined (see previous step), document the learnings, and update the initial business case. Define the next steps in the implementation roadmap.
	Deliverables: Big data pilot solution documentation, updated big data implementation road map and business case

Now with our basic knowledge, big data strategy can get started at an enterprise level, let's discover the building blocks of a big data solution. Figure 11-4 covers the key components of a big data solution. These include the following:

> Data integration engines—Ingest and extract data from numerous source systems (internal and external) and load it into information warehouses, operational data stores, and Hadoop based repositories. There are two types of data integration engines: 1) traditional data integration tools with big data integration capabilities (e.g., Informatica, InfoSphere DataStage, etc.) and 2) open source big data integration tools such as Talend Open Studio for Big Data or Pentaho Visual MapReduce. The key aspect is that these data integration tools have connectors to the Hadoop distributed file system (HDFS) or HBase (a NoSQL database optimized for use with Hadoop). Often in a complex big

data environment in which there is both a traditional information warehouse as well as a Hadoop repository, there is need for data exchange between the structured information warehouse and unstructured Hadoop repository. This is handled by Sqoop, a utility designed for handling bulk data transfers between the information warehouse and Hadoop. The Hadoop based integration jobs are scheduled using Oozie, an open source workflow and scheduling tool.

Shared data lake repository—As most enterprises have existing investments in information warehouses and operational data stores, the initial adoption of Hadoop has been to augment the capabilities of the existing information warehouse. This leads to a repository layer that is comprised of multiple repositories (information warehouse and Hadoop). This is often referred to as a shared data lake as it serves as a repository for all types of data— structured/unstructured, internal/external, traditional/new data sources. The data lake can be envisioned as a reservoir behind a dam where all the water is stored. This is an important distinction as the data in the lake is essentially data at rest. The other dimension of data in motion is not stored in a repository but ingested and analyzed using streaming integration technologies. Often for specific applications such as fraud analytics where there is a need to detect patterns in real time, stream-based integration is needed.

Often there is an analytical sandbox where the raw source data is stored for a period of time (depending on business needs). This repository is used for data exploration purposes and can be comprised of both structured and unstructured data. It can be considered as an extension of the Hadoop repository with a logical separation.

Information delivery and consumption—The information delivery and consumption layer is comprised of the different applications that consume the data stored in the data lake. These could be the data mining applications, business intelligence and analytical applications, and custom applications. Data visualization tools also help to provide interactive visualizations to end users and a seamless integrated view of both structured and semistructured data.

Information governance, quality, and metadata management—With the advent of big data in an enterprise, the paradigm of information governance, quality, and metadata assumes greater significance. With the advent of big data solutions, enterprises would retain data for longer periods as storage of large volumes in Hadoop is an optimal solution with lower costs. The challenge is created by new data sources such as social media data, sensor data, and the information governance strategy needs to decide who owns these data sets, who is responsible for their information quality, and who is responsible for their stewardship. With the advent of unstructured data, end-to-end metadata management is crucial to provide data lineage capabilities. Hence the choice of tools that integrate with the Hadoop platform and enable metadata integration is important from a solution architecture standpoint. Big data governance also needs to look at privacy and security issues about the new data types that come under its umbrella. In some countries there are regulations governing disclosure of these new data types, such as geolocation. When utilizing data monetization opportunities, enterprises also have to bear in mind what the customer data privacy laws are in a given country or state to avoid compliance issues when using customer location to provide customized offers. This means that there should be even greater emphasis on information governance and security.

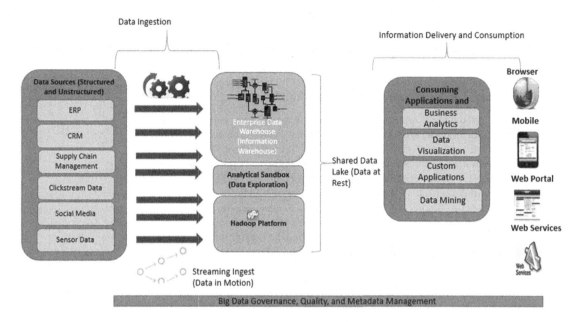

Figure 11-4. *Building blocks of big data layer*

The three key factors in the enablement big data solutions are—big vision and strategy, big data pilot and the next steps, and big data solution architecture and tools.

Big Data Vision and Strategy

Big vision and strategy originated as part of the initial assessment to gauge the current capabilities for the enablement of big data and the future road map based on the enterprise's vision. Big data vision is based on the enterprise's mission or vision statement and understanding how big data can help in realizing the vision. Big data strategy defines the big data use cases and initiatives that will help an enterprise to achieve its vision. In a close link to the strategy is the big data road map and business case, which help to determine the sequence of initiatives and the potential business benefits the big data use cases provide. In the case of big data the business case is more complicated to define as the return of investment (ROI) is difficult to obtain on the business use cases that have data monetization and revenue potential. The best case is to derive a ROI based on industry benchmarks if available or business process specific benchmarks.

Big Data Pilot and the Next Steps

Enterprises are still grappling with the adoption of big data solutions and developing the benefits in a structured manner. To ensure that the adoption of big data comes with the requisite benefits many enterprises start with a big data pilot project to assess the real benefits. This serves as a learning curve for the enterprises, as they go about fine tuning the use cases that can really benefit the enterprise in terms of revenue potential, new business models, cost optimization as well as enhanced customer service levels, and product innovations. For instance, if an enterprise wishes to integrate unstructured data into the customer data management landscape to provide unique insights concerning customer sentiments around products and specific brands; the steps involved in the proof of concept (PoC) would be as follows:

1. Identify potential use cases for social media integration and analysis (some of the common use cases include deep custom analysis on the topic of interest, build custom annotators for deeper analysis in terms of profession, interests, age groups, income levels, etc.).

2. Agree on specific use case and import data (e.g., Twitter, Facebook) for the given use case.

3. Prepare and extract relevant data for analysis with a data visualization/analytics tool.

4. Develop insights from both historical data as well as real-time data. Derive 360 degree view of customer from IVR (interactive voice response), web, social, and CRM data (see Figure 11-5).

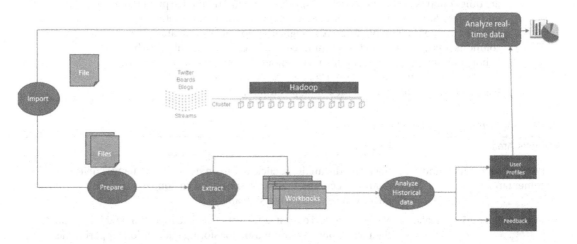

Figure 11-5. *Social media data integration proof of concept*

The benefits of a PoC approach ensure that the technology is proven before actual implementation. In addition there is a better understanding of the business benefits, for instance, in the of case social media data integration with customer data provides unique insights about acceptability of a product or brand that can help fine tune campaigns as well as manage customer service levels where there are frequent complaints. These provide better insights into customer retention and customer share of wallet that can help fine tune the use cases and the business benefits, which can help to define the next steps in the roadmap.

Big Data Solution Architecture and Tools

Key drivers in any big data solution are to define the solution architecture and the support tools and technologies. There are two key aspects to define the solution architecture: 1) identify the solution components needed to build a big data solution in a given enterprise and 2) identify the tools and technologies to build the big data solution. The key solution components that go into building a big data solution are as follows:

> Hadoop based repository—The big data solution must have a central repository for storing all types of source data (structured/semistructured/unstructured)— transaction data, enterprise data, machine data, social media data, and so forth. The Hadoop based repository stores data in the HDFS. Although the Apache Hadoop is an open source solution there are multiple adaptations of Hadoop distributions such as Cloudera, Hortonworks, and BigInsights. Hadoop is composed of the following modules: 1) storage based on HDFS, 2) resource management and scheduling to perform tasks, 3) distributed processing programming model based on MapReduce to support batch processing (there are other options now as well such as Apache Spark for interactive processing and Apache Storm for real-time processing), and 4) utilities and software libraries for the Hadoop platform. Although adoption of Hadoop and big data solutions was initially great in e-commerce and Internet companies, such as Yahoo and Google, it is now finding wider applications across many industries. Some of the common use cases of Hadoop are shown in Table 11-2. Also see Figure 11-6 for an overview of the Hadoop components.

Table 11-2. *Common Use Cases for Hadoop*

Analysis Area	Use Cases
Marketing optimization and customer targeting	Social media analysis, clickstream analysis, cross-channel behavior analysis, recommendation engines and customized targeting, campaign effectiveness, and conversion analysis
Enterprise data hub/data lake	Multistructured data staging, very fast data ingestion and integration, data warehouse augmentation for storing data for longer periods as well data warehouse off-loading, data exploration analysis on raw data, simple query, and reporting
Fraud detection and risk management	Network security monitoring, fraud behavior analysis, security, and event management
Operational analytics	Supply chain and logistics, smart meter analysis, well log analysis, asset management, and optimization

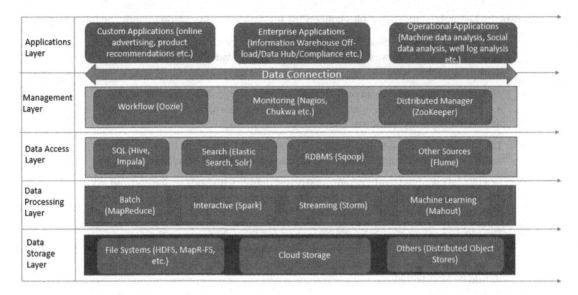

Figure 11-6. Hadoop ecosystem

Data ingest/integration tools—Data ingest/integration tools integrate structured and unstructured data from diverse data sources and load them into the HDFS or information warehouse depending on the big data solution architecture. In some instances there also is data exchange between the Hadoop and information warehouse; this is possible through the data integration tools, which must have the requisite connectors in the target system in question. Sqoop also can be used for this integration between Hadoop and information warehouse. The data integration engines can be traditional data integration providers such as Informatica or InfoSphere DataStage or open source data integration tools such as Pentaho or Talend.

Streaming integration tools—As discussed earlier data in motion has different challenges. It needs to be analyzed in real time using a stream-based integration approach, such as Apache Storm or InfoSphere Streams. Stream-based integration is used for real-time fraud detection, analysis of call drops, and other real-time operational analysis needs.

Data visualization tools—One of the key drivers for big data adoption is the ability to integrate and present both structured and unstructured data in the same report or dashboard. Big data visualization tools provide this ability that enables end users to access data on tablets and mobile devices as well as through portals. Data visualization tools such as Tableau, Datameer, and InfoSphere Data Explorer can provide rich visualizations that enable users to view data in widgets, graphics, infographs, and dashboards. Data visualization tools enable big data to be presented more intuitively.

Some of the key architecture principles concerning big data solutions would include the following:

- The big data solution should provide the organization with a trusted, unified, and consistent view of diverse data types integrated from a variety of structured and unstructured data sources.

- The big data solution should provide batch, interactive, and near real-time data integration and analytics capabilities. It must have the capability to handle mixed workloads and query patterns.

- The big data solution needs to scale well (to petabytes) to handle large volumes as well as to ensure that storage costs are kept low through HDFS based storage.

- In a big data enabled world where enterprises dabble with both structured and unstructured data, the big data solution requires that metadata be associated with both structured and unstructured data.

- The big data solution must have governance and the right to secure the data elements stored in its data lakes and to ensure there are suitable audit mechanisms. Compliance to local and global laws concerning data privacy and customer data should be maintained.

Along with the solution architecture definition, the optimal tools need to be selected to meet the big data requirements of an enterprise. Tool selection is based on a tool evaluation exercise. The solution outline phase is when the big data tool evaluations are done and the required tools are selected. Here I briefly cover the process of big data tool evaluation. There are a number of criteria based on which big data tools need to be evaluated. The evaluation criteria can be broadly classified into two areas namely

- Functional capabilities

- Vendor capabilities—Country support and product support

See Table 11-3 for big data tool evaluation criteria.

Table 11-3. *Big Data Tool Evaluation Criteria*

Evaluation Area	Evaluation Criteria
Functional: Hadoop based repository	Support for integration with existing information warehouse and other enterprise applications, ease of migration of data in and out of Hadoop, ability to scale and support variable workloads (batch, interactive, near real time), ability to provide insights on newly loaded data, support for business continuity and disaster recovery, high availability, and comprehensive administration utilities and security controls, support for open source, and Hadoop APIs (application programming interface)
Functional: Data integration/ingest tools	Support for data integration and ingestion from a wide variety of sources includes structured and unstructured data, connectors for HDFS and Hive, connectors for NoSQL and analytical databases (SAP HANA, Greenplum, etc.), parallel processing capability for high performance and scalability, easy to use graphical user interface for development, ability to generate code (Hadoop Pig Latin code), seamless integration of structured and unstructured data, application failover, multijob load balancing

(*continued*)

Table 11-3. (*continued*)

Evaluation Area	Evaluation Criteria
Functional: Stream integration	Support for processing massive amounts of streaming events (filter, aggregate, automate, rule, etc.), rapid integration with existing infrastructure and data sources, fast application development and deployment, live data discovery and monitoring, continuous query processing, alerting, connectivity with streaming data sources, support for fault tolerance and optimized performance
Functional: Data visualization tools	Support for key performance indicators, drill down, drill through, geospatial analysis, custom maps, interactive data discovery, visualization options, ease of use, printing, exporting, scheduled delivery, alerts, data sources supported, and many more
Vendor capabilities	Largest instance Different licensing modes and costs Collaborating with hardware and software vendors, integration with open source tools
Vendor: Product support	Training support provided Quality of product documentation Support for new upgrades and patches Availability of migration path from older to newer product versions
Vendor: Country specific support	Vendor presence in the implementation country Number of active clients and case studies Number of vendor consultants in the given country Number of consulting partners in the given country

Critical Success Factors in Big Data

Although embarking on a big data enterprise-wide program with a maturity assessment to understand current capabilities and gaps is a good starting point, a certain direction to sustain the momentum and realize the business benefits is also needed. In this section, the critical success factors in implementing a big data solution are covered.

As discussed in the section on building blocks, there is a sequence of activities performed while embarking on a big data program (see Figure 11-7).

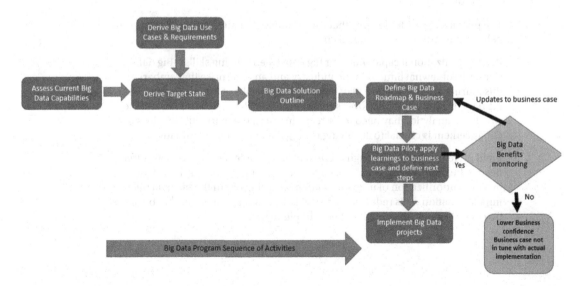

Figure 11-7. *Big data program sequence*

The sequence of activities include analyze current big data capabilities and derive big data use cases and requirements ➤ define target state ➤ big data solution outline (big data solution architecture and tools) ➤ define big data road map and business case (initial) ➤ big data pilot, apply learnings to business case and define next steps (the pilot gives a better understanding of actual benefits achieved and helps to define the next steps in the big data journey) ➤ implement big data initiatives that are defined based on the roadmap and pilot and then monitor the business benefits and feedback into a business case for a more realistic understanding of the business case. The key critical success factors are as follows:

Build on the roadmap based on use cases and business case—Two of the key deliverables of the initial big data strategy are the implementation a roadmap (with a list of big data use cases to support the enterprise business vision) and a business case. This is a good starting point, as at its core big data is about exploration of large data sets to provide insights that can be leveraged to transform the business (in terms of business models, revenue opportunities, cost rationalization, people engagement, product innovations, etc.).

Business alignment on big data uses cases—Another key difference between traditional information management projects and big data initiatives is that it is difficult to predict the business benefits up front based on a big data roadmap. The pragmatic approach is to define use cases in alignment with business stakeholders and to define the roadmap based on the prioritized use cases.

Start small with a big data pilot—Given the lack of experience enterprises face with big data initiatives, it is smart to start small with a pilot. Identify the top one to two use cases that are expected to provide value to the business and use them to implement the big data solution. This approach provides the following benefits:

1. It helps prove whether the technology solution works and integrates well with the existing information management landscape.

2. The actual benefits from the use cases implemented can be quantified and fed back into the initial business case to enhance business confidence in the business benefits.

The pilot also provides lessons that can be applied to the big data roadmap and help define further steps to be taken.

Build organizational capabilities for big data by enhancing skills—Big data skills are somewhat limited in the industry and an enterprise that embarks on this journey needs to have a plan to enhance the skills of the existing team in big data technologies. Moreover some specialized skills such as data science, streaming analytics may need to be recruited from the marketplace. The skills enhancement is crucial to achieving the big data road map of initiatives.

Build incrementally—Although there may be multiple big data use cases that need to be built, it is practical to design and build incrementally based on business prioritization of use cases and the learnings from the big data pilot implementation. This reduces the risk of failure and also ensures that business buy in is there for the initiatives being implemented.

Ensure information governance plays a vital role in the big data program—Big data initiatives involve integration of both structured and unstructured data and this brings new challenges concerning information quality, security, and ownership about the new data types being added to the information landscape. Therefore, information governance needs to be in place and the key roles identified within the enterprise to deal with these complexities. Chief Data Officers (CDOs) need to be involved with the big data programs to ensure that the right type of governance, compliance, and audit mechanisms are included as the enterprise moves on a journey of big data enablement. Governance is also important as metadata integration becomes more complex in big data solutions. Metadata from both traditional information management tools (data integration, relational databases, data modeling) and Hadoop needs to be integrated into the unified metadata repository for end-to-end data lineage capabilities.

Tools for Big Data

After reviewing the big data building blocks and critical success factors, I now cover some of the tools in the big data domain. Some of the key tools are Hadoop based repository, data integration/ingest tools, stream integration tools, and data visualization tools. The vendors covered have capabilities in all the areas mentioned. Some of the leading market tools are covered in Table 11-4.

Table 11-4. *Tools for Big Data Solution*

Big Data Components	Oracle	IBM	Open Source (Multiple Vendors)
Hadoop	Oracle Big Data Appliance	InfoSphere BigInsights	Cloudera, Hortonworks
Data integration/ingest	Oracle Data Integrator for Big Data	InfoSphere DataStage for Big Data	Talend Platform for Big Data, Pentaho
Stream integration	Oracle Complex Event Processing	InfoSphere Streams	Tibco StreamBase
Data visualization	Data Visualization for Oracle BI	InfoSphere Data Explorer	Lumify

■ ■ ■

CHAPTER 12

Building an Enterprise Information Management Solution

In Chapters 4 through 11, I discussed each of the key pillars that go into building an enterprise information management (EIM) solution. Although interrelationships between the pillars were discussed, for instance, in a master information management program there is significant dependence on information governance and quality, there is a need for end-to-end understanding of how the pillars need to be built to address the enterprise business vision. Although I covered the need for EIM reference architecture in Chapter 3, it is vitally important for an enterprise to have a program governance in place for all EIM initiatives. This helps to define the blueprint for EIM, which looks at both its support of the enterprise vision and business objectives as well as its adherence to the EIM reference architecture. This becomes more important with the advent of big data solutions resulting in more complex integration scenarios as well as the robust need for information governance and quality. As a lot of the unstructured data that enterprises now analyze are generated outside of firewalls, the need for a "big picture view" is essential to understand how these new data types influence the EIM reference architecture and the need for stronger program governance to address the objectives of the enterprise-wide EIM program.

Building an EIM solution not only needs adherence to the business objectives and developing business value but also needs to look at how this can be achieved within the given budget, resources, and skills in the enterprise. In this chapter the following topics are covered:

- *The key phases of an EIM program*

- *The need for EIM CoE (center of excellence)*

- *The delivery phases of an EIM project*

- *Critical success factors for EIM projects*

The following sections discuss each of these key considerations in more detail.

■ **Note**   The goal for this chapter is to explain the key phases of an enterprise information management (EIM) program, the need for an EIM center of excellence (CoE), the delivery phases of an EIM project, and the critical success factors that need to be factored during implementation of EIM projects.

Key Phases of an EIM Program

The need for program governance for managing EIM initiatives is being felt more and more in today's enterprises with their complex information needs. A recent survey by Forbes revealed the following:

- Data related problems cost the majority of companies more than $5 million annually. About 20% of enterprises lost more than $20 million per year.

- Ninety-five percent of enterprises agreed that robust EIM was critical for business success.

- Fragmented data ownership was the biggest road block to an EIM program.

- There is a need for closer alignment and communication between IT and business concerning EIM projects and their business benefits.

Insights from such surveys imply that enterprises need to have a robust blueprint for EIM both to manage costs and retain a competitive advantage as a business. Hence any enterprise that embarks on an EIM program needs to find an effective way to manage the program and measure its effectiveness. The key phases if an EIM program include the following:

- Build an EIM blueprint

- Define EIM program governance framework and organization

- Define EIM strategy and road map

- Implement EIM initiatives and measure benefits

Figure 12-1 illustrates the phases that constitute an EIM program.

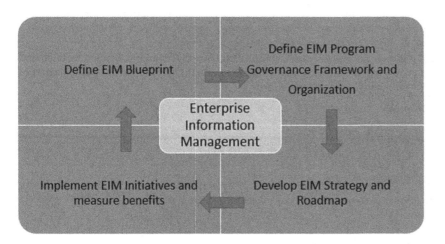

Figure 12-1. *Phases of an enterprise information management (EIM) Program*

Define EIM Blueprint

An EIM blueprint is a document that stipulates the key basic information management principles of an enterprise, and how these principles are developed from the business goals and objectives. The blueprint carries the business vision, which is mapped to the business goals and objectives which in turn are drivers for the information management principles. The information management principles are again broken down into EIM solution components that serve as enablers. Table 12-1 gives a few examples.

Table 12-1. *Business Vision Mapped to Business Objectives and Information Management Principles and Enablers*

Business Vision	Business Objectives	Information Management Principles	Information Management Enablers
Improve customer experience	Enhance customer service quality	Build 360 degree view of customer	Build customer master data management solution
Enhance operational efficiencies	Reduce operational costs	Reduce duplications in system interfaces	Build data integration hub
	Optimize operational processes	Near real-time operational reporting	Build near real-time operational data store (ODS) with change data capture (CDC)

As evident in Table 12-1, one of the business visions for the enterprise is to improve the customer experience. This can be achieved through numerous business outcomes, one of which is to enhance the quality of customer service. Customer service quality becomes a challenge as there are multiple touch points through which a customer interacts with an enterprise and consumes a product or a service. From an information management principle, it is vital to build a 360 degree view of the customer to gather insights about the customer as well as enhance the customer experience by learning from the challenges the customer currently faces. A customer master data management solution can help build a 360 degree view of customer across channels and serves as an information management enabler.

The EIM blueprint is constructed to keep in mind the business vision and objectives as well as the business architecture for the enterprise. The business architecture is the bridge between the business vision/strategy and the business functions at the tactical level. The EIM blueprint defines the EIM solution components that serve as enablers in achieving the business objectives and vision. The EIM blueprint also defines the interrelationships between the solution components and the order of execution in the overall reference architecture. The EIM blueprint is a working document and will need to be updated as the EIM practice matures in the organization. The EIM blueprint also should be updated once the EIM strategy and roadmap are defined.

Define EIM Program Governance Framework and Organization

Once the EIM blueprint is defined for an enterprise, there needs to be a governance framework in place to monitor progress and execution. The EIM program governance framework is defined at this stage to ensure that there are governance mechanisms in place to check the progress against the blueprint as well as provide overall governance to the EIM program. The EIM program governance organization is comprised of an ownership level that includes the CIO/COO (chief information officer/chief operating officer) who heads the information management function in the enterprise, the business sponsors (line of business heads), and the EIM governance lead. The strategic level is comprised of the EIM program manager and the EIM solution architects. The governance framework organization structure is shown in Figure 12-2.

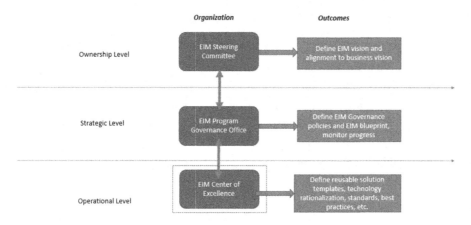

Figure 12-2. *Enterprise information management (EIM) program governance organization*

The operational level can be individual line of business IT teams (in the case of a decentralized model for information management project execution) or a center of excellence model (as shown in Figure 12-2). The EIM CoE model is gathering momentum as more enterprises are moving to this model. The section on the need for an EIM CoE later in this chapter covers this in greater detail. The respective organization levels and their outcomes are shown in Figure 12-2.

Develop EIM Strategy and Roadmap

Although an EIM blueprint provides the direction concerning the information management reference architecture and initiatives to support the business vision, a more detailed exercise would be to define an EIM strategy and roadmap, which typically breaks down the initiatives on a shorter time frame such as one to three years. EIM blueprints are the longer term views of where the EIM's program is headed, whereas the road map that was developed from the EIM strategy gives a shorter term view of the initiatives planned that was based on the business priorities and expected business benefits along with the ease of implementation.

EIM strategy provides a list of information management initiatives that need to be addressed to cover the current gaps in the information landscape. EIM strategy also defines a way for the information management function to have metrics to measure the effectiveness of information policies and processes to achieve business outcomes. One of the key outcomes of EIM strategy is the business case to measure the estimated benefits from the proposed initiatives.

Implement EIM Initiatives and Measure Benefits

Once the EIM strategy, roadmap, and business case are in place, the actual implementations can begin. The EIM projects are implemented based on the roadmap of initiatives and the business benefits from their implementations are used to update the business case.

Enterprises are increasingly looking toward the EIM CoE model to implement information management initiatives. This is the focus of the following section.

The Need for an EIM Center of Excellence (EIM CoE)

Once the EIM roadmap is defined and an enterprise wishes to implement the information management initiatives to deliver the expected business benefits, the organization needs to deliver the solutions as well manage costs, reuse solutions, and standardize solution designs and tools. This can be addressed through an EIM CoE. EIM CoE is a shared services function that can address enterprise-wide information management needs and provide fast, cost-effective deployment of information management projects by linking people, process, and technology across the enterprise. The key benefits of an EIM CoE are listed here and shown in Figure 12-3.

- Increased speed of delivery

- Cost optimization

- Create business value

- Create reusable assets

- Standardization of architecture and delivery models

- Project prioritization

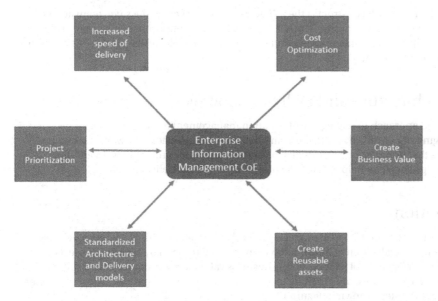

Figure 12-3. *Benefits from enterprise information management center of excellence (EIM CoE)*

Increased Speed of Delivery

One of the key benefits of an EIM CoE is the increased speed of delivery for information management projects. The increased speed of delivery is due to the following practices being leveraged: 1) use of standardized solution architecture and tools, which ensures that the existing solution building blocks are reused where feasible (there is no need to reinvent the wheel); 2) use of reusable solution assets built in different engagements such as ETL (extract, transform, and load) code snippets or report templates, ETL sequence designs, and document templates; 3) knowledge management, which ensures greater understanding of key performance indicators and measures through the use of business dictionary-like tools (as well as adoption if best practices such as design standards and coding standards); 4) reuse of EIM CoE team across projects; and 5) skills enhancement programs run by the CoE to enhance skills of existing team.

Cost Optimization

There are cost optimization opportunities that come with an EIM CoE organization model. These include 1) lower delivery costs due to reusable solution assets; 2) lower IT costs due to standardized architecture and tools; and 3) productivity benefits from the use of resources with experience in enterprise-wide information management projects and leveraging the existing knowledge of source systems, ETL, and BI solutions and data models.

Create Business Value

EIM CoE's focus on integrating data assets across enterprise applications as well as new age sources helps provide insights and knowledge about the enterprise's data assets and could leverage this insight across business initiatives to produce positive outcomes and better responsiveness to changing business conditions.

Create Reusable Assets

One of the key benefits of an EIM CoE construct is the inherent ability to leverage the same team across the spectrum of information management projects, which in turn creates the opportunity for reusable solution assets such as reference architecture, solution architecture templates, reusable design, and code (including ETL jobs snippets, report templates, etc.).

Standardized Architecture and Delivery Models

The benefit of a centralized approach to managing information management initiatives creates opportunities for leveraging standardized architecture and delivery models. These are well-documented tools and design patterns that bring desired results at a lower risk with better productivity. In addition knowledge from previous projects can help to fine tune the architecture and delivery models.

Project Prioritization

One of the processes introduced in an EIM CoE model is project prioritization. This is an often overlooked facet, but brings benefits in the realization of information management projects by ensuring projects are prioritized based on a combination of factors such as business value, service requirements, strategic alignment, ease of implementation, risk, and other factors. Project prioritization processes are part of setting up an EIM CoE and involve the steps shown in Figure 12-4.

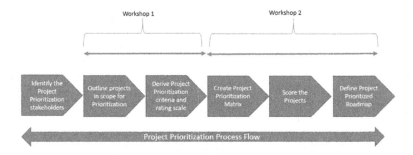

Figure 12-4. *Project prioritization process flow*

- Identify the project prioritization stakeholders—Identify the core project prioritization stakeholders for making decisions concerning agreement on the criteria for assessing the importance of projects as well as defining the roadmap based on the matrix designed.

- Outline projects in scope for prioritization—Once the budgets are defined for a given year, a workshop needs to be organized between the stakeholders to identify the projects in the entirety for the EIM CoE for the next 12 months. Outline the projects that need to be evaluated for prioritization, using the knowledge and expertise of the workshop participants.

- Determine project prioritization criteria and rating scale—The next step is to derive the factors that the stakeholders will use in assessing the relative importance of a project. Choose factors that help to differentiate a project from another, for example, the business value of a project, business priority, ease of implementation, associated risk, availability of key skills, availability of source data, infrastructure readiness, and so forth. Once the criteria are defined a rating scale needs to be designed on a scale of one to five that would be used to in assessing how a project would score. In addition to defining a rating scale, the rating criteria would need to be placed in descending order of importance and assigned a weight (see example below).

  - Required service: weight 5

  - Strategic alignment: weight 5

  - Business value: weight 4

  - Ease of implementation: weight 4

  - Availability of resources: weight 3

- Create the project prioritization matrix—Create the matrix by listing the criteria on the left, followed by weight, rating score, project score, and project weighted score.

- Score the projects—Score the projects based on the prioritization matrix rating criteria, and add the weighted values to determine the project's total score. If participant numbers allow, it is helpful to work in teams and to arrange for each project to be evaluated by two different teams. The benefits of this approach include: working in teams can produce more objective results because differing perspectives can be considered during the rating process. Although there are many projects to consider, dividing them among multiple teams helps speed the evaluation process.

- Discuss results and define prioritized project road map—After the projects have been scored, have a general discussion concerning the master list of prioritized projects. Note that the rating scores are an excellent way to begin discussions, yet still allow room for adjustment as needed. Remember that the prioritization matrix itself is just a tool, and the stakeholders scoring projects are using their best judgment. On review, the whole group may decide that a project needs to move up or down in priority, despite the score it received.

As a final step, a department or business unit may decide to establish groupings of projects based on natural breaks in scoring, for example high, medium, and low priority.

EIM CoE Organization Models

Some of the common EIM CoE models are covered in this section. The decision concerning which organization model to select for an enterprise should be based on the following factors: 1) enterprise culture concerning ownership of information management resources between business units and IT, 2) availability of specialized skills in the enterprise in the area of information management and analytics, and 3) prevailing model of centralized vs. decentralized IT delivery teams.

The common EIM CoE models are classified as 1) decentralized, 2) hybrid, and 3) centralized. The following is a comparison of the EIM CoE models.

1. Decentralized model—In this model the CoE is decentralized across business units with information management resources distributed across business units. The CoEs are local in nature and business unit specific. Although there may be some communication between the CoEs, it is highly unlikely that there would be any synergy between them in a decentralized model. In this model the CoEs are small in nature with specialized roles such as project management office (PMO) and solution architecture whereas the other development and support resources are with the line of business. CoEs are engaged with the proof of concepts whereas the actual delivery is performed by the line of business IT resources (see Figure 12-5).

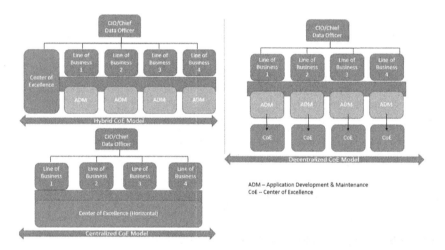

Figure 12-5. *Center of excellence (CoE) organization models*

2. Hybrid model—With this model there is a centralized CoE as well as CoE resources in line of business application development teams. The CoE here has more control in defining architecture and design standards and ensuring adherence to these standards. The CoE is not only involved in defining solution architecture and design best practices and POCs but also in the actual delivery of information management and analytics projects. CoEs provide the architecture and designer skills in implementation projects. This ensures there is better adherence to design and architecture standards as well as the use of reusable design patterns and solution templates. The hybrid model is more mature than a decentralized CoE model. In this model the CoE has limited control over the application development resources in terms of defining their project involvements and learning of new skills.

3. Centralized model—In this model all information management initiatives are handled by the CoE. The CoE serves as a horizontal by catering to all information management needs across the line of businesses (LOBs). All implementation resources are within the CoE organization, which gives it flexibility in controlling the resource allocation to projects based on skill needs. In addition the CoE can define its own skill plan for resources based on annual projection of skills needs. This projection comes in through project prioritization of projects to be implemented over a time period of 12 to18 months. The CoE is more mature in this model with well-defined processes, estimation templates, solution architecture patterns, and design best practices. The CoE works in close collaboration with the LOBs and has representation on the CoE governance organization to ensure that the projects implemented are within the parameters of business value, business priority, and available budgets.

With an understanding of the CoE organization models I now look at what constitutes an EIM CoE. The building blocks of an EIM CoE are comprised of 1) a governance layer, 2) delivery and maintenance services, and 3) competencies and capabilities. These layers are represented in Figure 12-6.

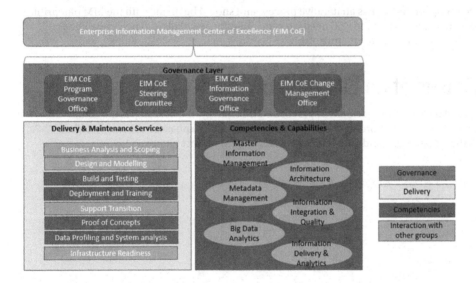

Figure 12-6. *Enterprise information management center of excellence (EIM CoE) layers*

1. Governance layer—The governance layer of the EIM CoE is comprised of the EIM CoE program governance office, the EIM CoE steering committee, the EIM CoE information governance office, and the EIM CoE change management office. The program governance office is responsible for the overall management of the EIM program, resource tracking, and financial management. The steering committee includes the CIO/chief data officer as well as a group of business executives that help steer the strategic direction of the EIM CoE in close alignment with business vision. They also help identify the stakeholders who would define the prioritization of projects to be executed by the EIM CoE in the given period. The information governance is comprised of the chief data officer and lead data steward along with the group of business representatives that help define the information governance charter and align resources for defining the information governance policies. The change management office is responsible for all change management activities related to the projects implemented by the EIM CoE.

2. Delivery and maintenance services—The delivery and maintenance services offered by a mature CoE include business analysis and requirements scoping, design and modelling skills, delivery and testing services, deployment and training of end users, and support transition from the development team to support desk. In additional the CoE provides services in data profiling and source system analysis for new initiatives, proof of concepts concerning new technologies to prove that the technology can provide the desired benefits, infrastructure readiness assessments at the start of new projects or initiatives in conjunction with the infrastructure teams to ensure that the infrastructure build happens at the right time based on the implementation road map.

3. Competencies and capabilities—One of the core functions of the EIM CoE is to provide specialized skills in the information management disciplines it supports including master information management, information integration and quality, information architecture, metadata management, and so forth. The competency building focusses on new skills enhancement, knowledge management, building best practices, and adherence to best practices.

The journey to set up an EIM CoE is an iterative process and should be in line with the EIM blueprint. As the projects become implemented the CoE needs to keep building new skills and new capabilities to transform them from a start-up CoE to a mature one.

Delivery Phases of an EIM Project

Once the roadmap and business case are built as part of an EIM strategy, the next step is the actual implementation. The EIM project or program can be executed in multiple phases as shown in Figure 12-7. Table 12-2 defines the key activities and deliverables for each of EIM project phases.

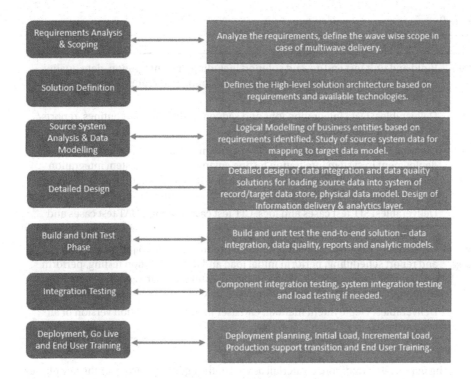

Figure 12-7. *Enterprise information management (EIM) project phases*

Table 12-2. *Activities and Deliverables for EIM Project Phases*

Engagement Phase	Activities and Deliverables
Requirements analysis and scoping	Requirements analysis and scoping for releases by interviewing business stakeholders. Requirements analysis included business definitions of measures, KPIs, look and feel of reports/dashboards and visualization methods to be used.
	Deliverables: Business requirements documents (BRD, data visualization strategy)
Solution definition	High-level solution architecture based on requirements, available tools. This includes data sourcing and integration, information quality, information delivery and consumption.
	Deliverables: Solution architecture document
Source system analysis and data modelling	Once the requirements are finalized, the logical data model is designed and the source system analysis takes place to map the logical data model entities to source system attributes. The transformation rules are defined and data granularity is looked at, which in turn defines the aggregation needs if any.
	Deliverables: Logical data model and source to target mapping
Detailed design	Detailed design of data integration jobs and sequences, data quality routines and information delivery reports/dashboards. Physical data model design based on target data store.
	Deliverables: High-level and low-level design documents

(continued)

Table 12-2. (*continued*)

Engagement Phase	Activities and Deliverables
Build and unit testing	Build and unit test end-to-end solution including data integration, data quality, reports/dashboards. Set up and test job scheduling and report bursting based on needs.
	Deliverables: Data integration jobs and sequence, data quality routines, reports/ dashboards, data model
System, component integration testing and user acceptance testing	Component integration testing (CIT) of all components of the solution, code deployment in system integration testing environment and system integration testing. Fix defects and again deploy code in system integration testing (SIT) for another cycle of testing. Perform user acceptance testing (UAT) and fix defects.
	Deliverables: SIT test cases and logs, CIT test case and logs, UAT test cases and test logs
Deployment, go live, and end user training	Deployment planning, code migration to production, set up job scheduling and report scheduling, perform initial load and data validation testing, perform catch up incremental loads, data validation testing. Prepare go live system communications and end user manuals. Provide IT and business user training.
	Deliverables: Deployment manual, end user manual, production version of all code artifacts

EIM programs can be managed in traditional waterfall as well in the agile way. However the key phases discussed here have to be executed to ensure that the deliverable quality is of optimum standard. When managing large programs, upfront planning and due diligence needs to be done to determine where the solution needs to be introduced first (in the case of multicountry rollout) and which part of the solution needs to be implemented first. Here again the EIM blueprint and strategy can serve as pointers and a due diligence performed on the countries/markets in scope helps decide the order of execution. The project prioritization matrix can be looked on as useful tool to determine the relative importance of projects, the order of execution, and how business benefits can be delivered in a timely manner to increase business confidence on the EIM solution.

One key challenge in large multicountry rollouts of EIM solutions (i.e., information warehouse, master information management, etc.) is the order of execution of the solutions as well as the need to cater to both global and market specific local requirements. As part of the EIM program planning it is important to define what constitutes a global and local solution; of course, the global solution needs to be deployed in all markets/countries and the intricacies of the local solution need to be determined from one country to another. Standardization of solution architecture, data integration design and data models does help to bring consistency to the solution as well as reduce the time to market. In the next section the critical success factors involved in the execution of an EIM program are discussed.

Critical Success Factors of an EIM Project

Although you have embarked on a large EIM program with a maturity assessment, there is a need to sustain the momentum and realize the business benefits. In this section, the critical success factors in implementing an EIM project/program are covered.

The EIM program starts with an assessment of current capabilities that involves reviewing the EIM blueprint and deriving the key EIM requirements from a program standpoint (see Figure 12-8).

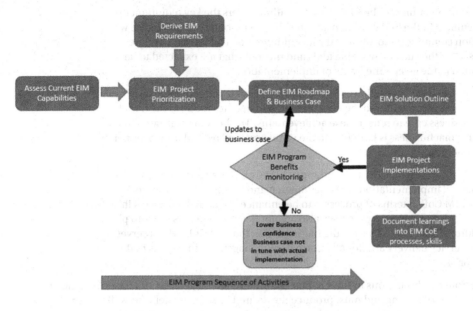

Figure 12-8. *Enterprise information management (EIM) program sequence*

The sequence of activities shown in Figure 12-8 is summarized in the following:

Assess current EIM capabilities—Once the EIM blueprint has been defined for the organization based on the business vision and business objectives and the EIM program organization is in place, the first step is to look at current EIM capabilities in terms of functional coverage as well as maturity of the information management practices. This helps to determine if there are gaps in the current landscape with respect to the target state as defined by the maturity assessment.

Derive EIM requirements—Parallel to the EIM capabilities assessment is the process of developing EIM requirements. The EIM requirements are derived based on EIM blueprint as well as understanding the current EIM requirements based on future business vision and objectives. The EIM requirements need to be defined as information management initiatives to support the targeted business objectives as these could be further crystalized as projects. The list of initiatives/ projects could serve as input for the project prioritization exercise.

EIM project prioritization—In this step the projects are rated based on the project prioritization matrix as defined by the core stakeholders based on criteria such as strategic alignment, business value, ease of implementation, availability of resource and requisite skills, availability of source system data, and so forth. The projects are ranked based on the project prioritization matrix and follow-up workshops can serve as input into the EIM roadmap.

EIM roadmap and business case—Once the list of prioritized projects is defined by the core stakeholders, the EIM roadmap is available for a given period of time (12 to 24 months) and the initial business case is defined based on the top priority projects that are to be implemented in a given time frame (next 6 to 8 months). This provides a tangible understanding of the expected short-term business value.

EIM solution outline—The EIM solution outline covers the key requirements as identified in the EIM roadmap and the high-level solution architecture with solution components to address these requirements. The solution outline is based on the current available tools and the tools that are expected to be acquired in the given time frame of implementation.

EIM project implementations—The actual projects are implemented and the business benefits achieved are documented and updates made to the initial EIM business case to reflect these achievements. The benefits realization focus ensures that business is bought into the value provided by EIM initiatives being implemented.

Document knowledge into the working of EIM CoE—The key knowledge from the project implementations are documented and inputs fed into the working of the EIM CoE in terms of processes to be enhanced or added. Skill gaps in technologies or business processes also are notified and a plan is put into place to address these gaps as part of the maturation in the EIM CoE, which serves as the service delivery arm for all information management initiatives in the enterprise.

As is evident from the discussions in this chapter, information management initiatives often need business involvement and funding and must produce the desired business benefits as well as engage business stakeholders. The key critical success factors for an EIM project/program are as follows:

Business alignment on EIM roadmap—One of the key success factors in an EIM program is to ensure the correct business participation from the time of the definition of the EIM blueprint, the EIM program governance framework and organization, and the EIM roadmap and business case. Business participation during the implementation phase will ensure a wider adoption of the EIM solutions, which will provide greater business value through analytical decision making and information sharing among business units.

Start small and build incrementally—Given the lack of expertise some enterprises face with EIM initiatives, it is good to start small with a smaller set of projects. Identify the top one to two projects that are expected to provide value to the business and then implement the EIM solution. This approach provides the following benefits: 1) faster time to market for the realization of business benefits, 2) updates to the business case ensures greater understanding of the benefits delivered, and 3) reduces the delivery risk of large, complex EIM programs where delivery is broken into sections also makes it easier from a change management perspective.

Build organizational capabilities for EIM by enhancing skills—Some EIM skills such as information governance, information quality, and big data are somewhat limited in the industry and an enterprise that embarks on this journey needs to have a plan to enhance the skills of the existing team in these specific processes and technologies. The skills enhancement is important to achieve the roadmap of initiatives in a timely manner as well as in the maturation of the EIM CoE.

Ensure information governance plays a vital role in the EIM program—EIM initiatives with the advent of big data integration, involve integration of both structured and unstructured data and this brings new challenges concerning information quality, information security, and ownership about the new data types that are being added to the information landscape. Therefore, information governance needs to be in place and the key roles identified within the enterprise to deal with these complexities. Chief data officers need to be involved with the EIM programs to ensure that the right type of governance, compliance, and audit mechanisms are built in as the enterprise moves on a journey of EIM enablement. Information governance plays a pivotal role in master information management, single view of customer, and customer service quality enhancement type of initiatives. The earlier information governance is introduced in the lifecycle of an EIM program, the ownership of the data assets will be more effective and the quality of information processed through the EIM solutions will be improved. Better information quality means better quality of business decision making.

■ ■ ■

CHAPTER 13

EIM in Today's Business Environment

In Chapters 4 through 11, I discussed the key pillars that built an enterprise information management (EIM) solution, and how an enterprise went about building an EIM solution. In this chapter, I focus on some of the key trends that are emerging in the EIM landscape. Enterprises are increasingly faced with new challenges to stay ahead of the competition and this has resulted in many transformations and disruptions in the enterprise landscape. I also discuss big data use cases from the oil and gas, retail, and mining industries. In this chapter the following topics are covered:

- *The key EIM trends in today's enterprises*

- *Big data use case for oil and gas*

- *Big data use case for mining and metals*

- *Big data use case for retail*

- *Emergence of cloud-based EIM solutions*

In the next few sections I discuss each of these key considerations in more detail.

■ **Note** The chapter goal is to explain the key trends that are emerging in the EIM landscape of large and small enterprises; some big data use cases from the oil and gas, mining, and retail industries; and the emergence of cloud-based EIM solutions as a new transformation enabler.

Key EIM Trends in Today's Enterprises

The ever increasing need for timely data in decision making and as a strategic differentiator has made information management and analytics a top priority for chief information officers (CIOs) as well as chief strategy officers (CSOs). However the ever increasing complexities in the business environment are changing the landscape of EIM solutions such as the factors given in the following:

- The need for information sharing in global supply chains with trading partners and suppliers.

- The emergence of big data needs (high volume, new varieties of data, high speed of processing).

- Lowering costs of information management.

- The rising needs of data privacy, security, and compliance report.

- The emergence of social media data as providing unique insights about customer decision making.

These key trends emerging are summarized in Figure 13-1.

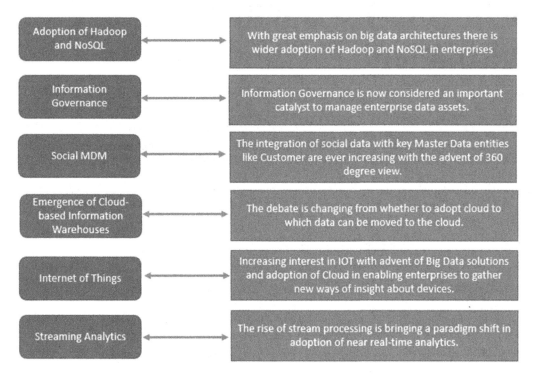

Figure 13-1. *Key enterprise information management (EIM) trends in today's enterprises*

I look at each of the trends that are bringing change in the EIM solutions landscape in the following sections.

Adoption of Hadoop and NoSQL

With more enterprises moving on the big data bandwagon, there is a rise of Hadoop and NoSQL (nonrelational) in the EIM architecture. One of the common use cases in the early adoptions of Hadoop are enterprises building a data lake for hosting both raw and processed data as well as retaining data for extensive periods of time for both analytic and compliance needs. With a strong business case concerning the ability to store both semistructured and unstructured data at much lower costs and the ability to retain this data for longer periods of time, the adoption of Hadoop has seen a significant upsurge in recent years. One of the key drivers behind the rise of NoSQL databases (not only SQL) is that it enables developers to create new applications or modify existing ones much faster than any of the legacy RDBMS (relational database management system) would permit. However one challenge that remains with the NoSQL databases is the lack of a quality analytics toolkit to leverage insights into the data. (NoSQL data is not tabular or uniform and cannot be leveraged using existing analytical toolkits).

Information Governance

With the rise of big data solutions as well the emergence of the cloud, the need for governance of enterprise data assets has never been higher than it is today. With the emergence of global supply chains, more data exchange is taking place between trading partners and suppliers and the need for governance is significant. In a big data enabled enterprise, data can be structured or unstructured and generated from both internal as well as external sources. This enhances the need for governance of data assets, suitable ownership of data assets, and information quality policies in place to ensure that the data assets are trustworthy. With greater emphasis on compliance reporting, information governance and risk management are now being embraced as crucial differentiators by enterprises.

Social Master Data Management

With the advent of social media and customers using this medium to discuss brands, product feedback, and complaints, the importance of social media data has increased in leaps and bounds. When enterprises discuss the 360 degree view of customers and products, it is impossible to achieve without integrating unstructured data concerning customers and products along with the structured master data. This is the future for enterprises who seek deeper insights from social media data concerning customer feedback about products and services as well finding the right products to sell to the appropriate set of customers. With the advent of big data solutions the integration of structured and unstructured data can be managed, however data quality remains an area that requires governance and controls.

Emergence of Cloud-Based Information Warehouses

With the increasing costs of large corporate information warehouses (especially for transnational corporations with numerous country specific warehouses to support decision making in additional to global information warehouses), there is a paradigm shift within enterprises looking at leveraging clouds to deploy large warehouses. There are two benefits: 1) moving to a pay-per-use model that is usually prevalent in cloud-based deployments and 2) reducing storage costs. The advent of Amazon's Redshift brings cost benefits in comparison with traditional approaches to information warehouses. The cost per terabyte of storage is significant lower in cloud-based models, which allows the enterprise to offer their information warehouse as a service (data as a service). There is growing competition in this area with IBM's dashDB Enterprise MPP running on Bluemix platform services and Microsoft's Azure SQL Data Warehouse.

Internet of Things

Internet of things (IOT) is creating a buzz as more enterprises become instrumented and interconnected. IOT can be visualized as a set of devices registered through an IOT platform that continuously emits data through sensors. The massive amounts of data generated through these devices would need to be analyzed in near real time to gain insights about the workings of business processes, machine efficiencies, potential events, and breakdowns. These insights would provide opportunities to further optimize the business processes and asset maintenance processes making businesses smarter. It would impact people's lives as well as the devices that they use (e.g., TVs, cars, refrigerators, etc.).

Streaming Analytics

A key transformational technologies is stream processing. Stream processing is a game changer in the big data enabled world as it enables data to be analyzed in motion as compared to traditional methods such as change data capture, which involved latencies in terms of changed data being stored and analyzed. Stream processing enables enterprises to react to changing business conditions in near real time and that enables applications such as fraud detection, trading, and system monitoring. Stream processing is when data streams or sensor data is processed (high-event throughput versus number of queries) while complex event processing (CEP) utilizes event-by-event processing and aggregation with business rules and logic. In contrast to traditional RDBMS models in which data is first stored and then subsequently analyzed through queries, stream processing analyzes data while it is in motion. Stream processing also can connect to external data sources enabling applications to incorporate selected data into an application flow or update an external database with processed information.

Impact Areas

Here are a few examples of potential impact areas for each of the key EIM trends as shown in Table 13-1.

Table 13-1. *Key EIM Trends and Associated Use Cases*

Key EIM trends	Potential Impact Areas
Adoption of Hadoop and NoSQL	Big data integration Cost optimization, etc.
Information governance	Single view of enterprise, risk management and compliance, etc.
Social master data management	360 degree view of customer, products, enhance customer service, customer product affinity, etc.
Cloud-based information warehouses	Data as a service, cost optimization, etc.
Internet of things	Preventive maintenance, environment monitoring, smarter utilities, etc.
Streaming analytics	Fraud detection, Network monitoring, Risk management, etc.

As shown in Table 13-1, there are numerous impact areas for each of these key EIM trends. I now delve deeper into some big data use cases by industry. Although there are numerous case studies for the telecom, health care, banking, and financial services, the focus of the use cases in this chapter will be some of the later adoptions to big data solutions.

Big Data Use Case in Oil and Gas

Before looking at the common big data use cases in the oil and gas industry, I analyzed the value chain for a better understanding of the optimization opportunities. The oil and gas value chain is shown in Figure 13-2 and consists of the following steps:

- Exploration
- Production
- Processing
- Storage
- Transportation
- Refining
- Distribution
- Marketing

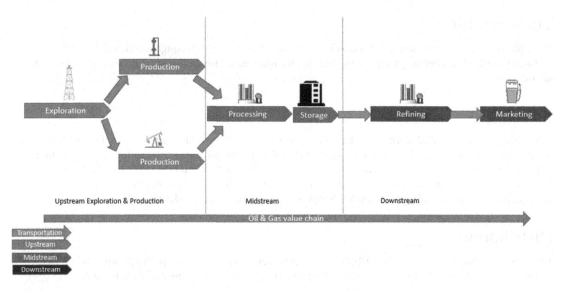

Figure 13-2. *Oil and gas value chain*

Exploration

During exploration, the oil and gas conglomerates drill exploration wells. Huge volumes of data (geological, seismological) are created as well. These exploratory data sets are fed into computer models to generate reservoir potential and geometry. As exploration is a costly exercise with high risks, reservoir potential may not be economically viable in many instances.

Production

Once a reservoir is considered economically viable, which is based on the exploration data and surveys; production begins. As drilling starts and the production drills run on full capacity; equipment maintenance is important to ensure there are no outages. There are numerous challenges faced during drilling including gas flares, reservoir outputs not up to par with expectations, oil recovery rates going down with reservoir ageing, and so forth. A huge amount of production data is created through well logging, and the data provide insights into the reservoir properties and the monitoring of the drilling equipment.

Processing

Some basic processing is done by midstream companies, including natural gas processing that purifies raw natural gas, removing sulphur, and producing natural gas liquids.

Storage

The oil depot in the oil refinery is used to store the bulk liquid products. There are also storage depots to store finished products.

Transportation

Once the crude has been extracted, it has to be transported to refineries through a series of networks including crude oil carrier ships, pipelines, and surface transport. The crude is finally moved to refineries for further processing.

Refining

One of the key value-added activities is the refining process through which crude is refined to produce fuels of different types and useful by-products. Crude oil is refined in an oil refinery to produce products such as naphtha, gasoline, diesel fuel, heating oil, kerosene, liquefied petroleum gas (LPG), lubricating oils, and some by-products, which include hydrogen, light hydrocarbons, and pyrolysis gasoline. Oil plants also have an oil depot to store incoming crude oil feedstock as well as bulk liquid products.

Distribution

The finished products need to be distributed to different retail outlets such as petrol pumps and service stations. The distribution includes transportation by railway, roads, shipping, and river barges.

Marketing

The finished products are marketed and transported to large retail outlets such as petrol pumps and service stations.

Potential Use Cases

At each stage of the value chain, data is created and insights can optimize the operations. See Table 13-2 for examples of big data use cases at each stage of the value chain.

Table 13-2. *Oil and Gas Big Data Use Cases by Value Chain*

Value Chain Phase	Potential Use Case
Upstream: Oil exploration	Identifying new oil prospects: Analysis of well data, seismic data, and industry trends can be used to evaluate potential oil fields. This also can be used to identify premium oil drilling locations and bid for prospecting licenses.
Upstream: Production	Enhance oil recovery and predict oil production levels: Analysis of seismic data, drilling data, and reservoir data is crucial to enhance oil recovery from existing oil wells as the cost of drilling new ones is high. This analysis helps to determine the oil lifting methods. In addition analysis of data from production well can help in predicting the production levels in oil wells, which is crucial for investor confidence.
Upstream: Production	Asset maintenance and monitoring: Potential drilling issues and equipment outages can be understood by analyzing sensor data from equipment (such as drill heads, bore hole camera sensors, etc.) as well as geological data such as overburden thickness and rock type. Analysis of equipment data in near real time can help in planning preventative maintenance and ensure minimal drilling outage and downtime.

(*continued*)

Table 13-2. (*continued*)

Value Chain Phase	Potential Use Case
Upstream: Production	Safety and environment: By analyzing data from multiple sources such as reservoir geology, drilling equipment, weather data drilling issues, and environmental risks can be predicted and drills can be proactively shutdown preventing financial losses as well as reputational risk.
Downstream: Marketing	Analyze large volumes of customer data rapidly to identify buying patterns to identify new cross-sell opportunities. Segment customers based on buying propensity for finished products as well as by-products.
Downstream: Refining	Preventive maintenance: Real-time equipment sensor monitoring systems in refineries that analyzes machine data and optimizes the maintenance schedule based on refinery productions needs and equipment health indicators.

These are some of the common use cases for adoption of big data in oil and gas industry. There are other use cases as well to meet technical considerations such as data storage needs for seismic data from reservoirs. A traditional challenge was getting the seismic data on tapes to the data centers for analysis resulting in lack of real-time or near real-time analysis capabilities. With the advent of Hadoop and big data this is fast changing.

Big Data Use Case in Mining and Metals

Currently companies in the natural resources industry spend just 1% on IT as compared to 6 to 7% for most other industries. This low investment traditionally means that mining companies collect data but do not utilize the data sets to their fullest. Big data solutions have the ability to transform mining companies by leveraging the data sets across the value chain and help the companies gain a competitive advantage.

Some of the key information management challenges faced by the mining companies today are summarized in the following:

- Diverse types of data from production drills, mine planning systems, GIS data sets, project schedules, equipment monitoring, mineral extraction, and processing.

- The continuous need to increase productivity requires an integral, real-time overview of the production process.

- The inclusion of new equipment demands rapid integration of their operational statistics.

- The ability to handle production issues in a timely manner based on real-time views of production processes such as drilling, equipment monitoring, and so forth.

- Manage performance from multiple perspectives such as managing operational efficiency, health, safety, and environment, regulatory needs, supply chain leads to an integrated information management capability.

- Ability to handle real-time data feeds such as drilling data, equipment utilization and monitoring as well as data at rest from enterprise applications such as ERP (enterprise resource planning).

All these challenges can be met by the adoption of big data solutions. Before embarking on the big data use cases in mining, the mining value chain is discussed in Figure 13-3.

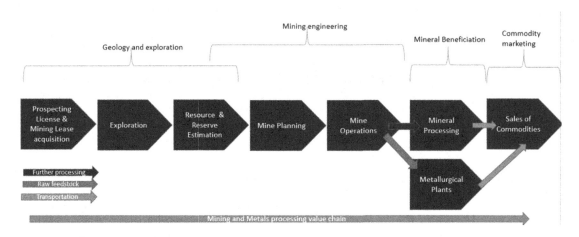

Figure 13-3. *Mining industry value chain*

Prospecting License and Mining Lease Acquisition

The first step for a mining conglomerate is to apply for a prospecting license based on the business plan. If the prospecting lease is in an area that has a suitable potential, a mining lease also is acquired for mining operations.

Exploration

Once the mining conglomerate has the prospecting lease, it can begin the geological exploration, which involves surveys, geological mapping, and exploratory drilling. The exploration data are analyzed to understand the ore orientation and quality based on samples collected.

Resource and Reserve Estimation

Results of the exploration are fed into the computer models to derive the resource and reserve estimation. At this stage the mining conglomerate has to decide whether the reserve is economically viable to mine. If not already acquired, the company needs to apply for a mining lease to begin.

Mine Planning

Before the mining can commence, one crucial activity is needed to determine the mine plan. The mine plans vary depending on the method of mining—open cut vs. underground as well as based on the mineral type, ore body thickness, and orientation. Mine plans are made that keep in mind the mine layout, equipment selection, ore reserves, and rate of production expected, cost estimation, and economic evaluation and sustainability.

Mine Operations

The key mine operation activities include 1) excavating earth and rock (this includes blasting); 2) processing of the excavated material to separate the ore from the waste rock; 3) storing waste material in dumps; 4) monitoring of environmental conditions such as air quality, water quality, and noise levels; and 5) supporting services such as maintenance of equipment repair shops, labs, offices, warehouses, and so forth.

Mineral Processing

Some minerals are sent for further processing. Mineral processing is the separation of commercially viable minerals from their ores. Mineral processing also is known as ore dressing. Mineral processing can involve four types of operations: 1) comminution, which involves particle size reduction; 2) sizing, which involves the separation of particle sizes by screening or classification; 3) concentration, which increases in percentages of mineral in the concentrate; and 4) dewatering, which involves the solid/liquid separation.

Metallurgical Operations

Although some minerals are sent for further processing, others such as iron ore, are sent as feedstock for metallurgical operations. Metallurgy involves the processing of ores to extract the metal and the mixture of metals with others elements to produce alloys.

Transportation

Transportation occurs in three places 1) from the mine benches or tunnels to the mine operations processing plant, 2) from the mine operations plant to either the mineral processing unit or metallurgical plant, and 3) the finished products shipped to the point of sales such as industrial consumers and wholesalers.

Sales of Commodities

Once the finished products such as steel ingots and steel pipes are produced they are sold to industrial consumers and wholesalers.

Potential Use Cases

Table 13-3 defines the main big data use cases for the mining and mineral processing industry.

Table 13-3. *Big Data Use Cases for the Mining and Metals Industry*

Value Chain Phase	Potential Use Cases
Geology and exploration	Efficient exploration and drilling: Analysis of geological data and drill log analysis helps companies perform reserve estimation in a more timely and accurate manner. This also helps in planning future bore holes based on modelling the ore body layouts. This can help reduce costs by focusing on drilling in more potential areas and can help reduce the exploration risks as well.
Mine planning	Design for enhanced ore recovery: Analyzing geotechnical data from exploration drills can help mine planners design for steeper batter angles that add value to an ore resource by reducing overburden removal and access to greater percentage of the ore body.
Mine operations	Transportation optimization: With modern dumpers equipped with GPS and an ability to analyze movement of dumpers in real time, mining managers can optimize transportation of blasted ore from the mine benches and underground tunnels in an efficient manner.

(*continued*)

Table 13-3. (*continued*)

Value Chain Phase	Potential Use Cases
Mine operations	Avoid production delays and safety hazards: Leverage geotechnical data such as joint planes, faults to update geological models of a mine that help to avoid geotechnical hazards such as wall collapses, shovel and crane operations, as well as production delays.
Geology and exploration, mine operations	Internet of things (IOT): IOT enables shift managers to analyze drill bit data in real time, which helps to plan for replacements and avoids down time due to drill bit wearing. IOT enables mining companies to build connected mines that are digitally integrated and provide a real-time view of operations from end-to-end. This enables mining operations to run efficiently at a lower cost with lower risks of down time.
Mineral processing	Reduce power consumption and maintenance costs in mineral plants: Analyzing acoustic data from sensors in the mill chambers gives insights into distribution of steel balls in the mill that grind ore, which has an impact on power costs as well as wear and tear of mill equipment.
Mine Operations	Predictive maintenance: By analyzing real-time sensor data, companies can get deep insight about when a machine or a part needs repair or replacement. Companies can project based on historical data, forward forecasts, and algorithms, when machines are likely to break or need service and plan for predictive maintenance at much lower costs.

Rio Tinto's Mine of the Future program includes setting up a digital mine where operations are interconnected and there are real-time insights into each component of the mining operations. Part of the Mine of the Future program is setting up an analytics center of excellence (ACE) that will analyze equipment productivity across its global operations. The ACE will analyze massive amounts of data captured by an array of sensors attached to fixed and mobile equipment and enable experts to predict and prevent engine breakdowns and down time events, thereby boosting productivity and ensuring safety.

Big Data Use Case in Retail

With traditional brick and mortar retailers losing sales to e-retailers, there is a greater emphasis on big data analytics as a differentiator to understand consumer behavior as well as look at cost optimization opportunities. According to a survey by Gartner in 2013, retail big data analytics market will touch 4.5 billion dollars by 2019. The big data challenges that retailers face today include the following:

1. Challenges in predicting customer buying habits—Customers have been spoiled for choice in terms of ways of delivery as well as varieties of products. With the emergence of mobile technologies, customers can search for products and interact with peers over social media to seek feedback about products and services.

2. Promotions and campaigns not delivering the expected returns—Customers may take note of a campaign or promotion, but may choose to ignore it. Retailers are finding it increasingly difficult to understand customer behavior and brand connect.

3. Optimal pricing—With the emergence of e-retailers and an efficient supply chain, product pricing has become very complicated and is often the deal breaker. Customers are extremely well informed through technology and social media and are quick to understand the suitability of prices for a given set of features.

4. Fraud—The increasing incidents of fraud, such as the fraudulent return of products or credit card information theft, result in operational risks and reputation challenges for retailers.

All these challenges can be met by adoption of big data solutions. Before embarking on the big data use cases in retail, the retail value chain is discussed in Figure 13-4 and in the following sections. At the two ends of the retail supply chain lie the suppliers to the left who supply the requisite products and customers at the right who consume the products and services.

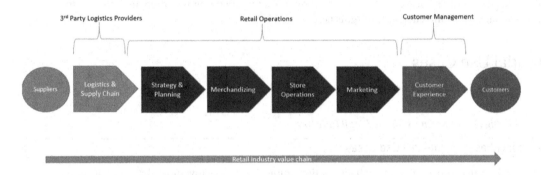

Figure 13-4. *Retail industry value chain*

Logistics and Supply Chain

The procurement of products and the management of suitable inventory levels are kept by third-party logistics providers. The key focus areas in this area are supplier performance management, inventory management, and logistics management.

Strategy and Planning

This is one of core capabilities of a retailer with a focus on real estate planning; sales planning; and forecasting, competition analysis, and benchmarking.

Merchandizing

This is another of the core capabilities of a retailer with a focus on assortment planning, markdown optimization, price optimization analysis, market basket analysis, cross-selling opportunities, and competitive intelligence.

Store Operations

Store operations, another core capability of a retailer, has its focus on the following activities: layout optimization, labor optimization, loss prevention, and store profitability analysis.

Marketing

Marketing is a core retailer function with a focus on the following activities: customer segmentation, cross-channel synergies, marketing effectiveness, and promotion effectiveness.

Customer Experience

One of the key differentiators for retailers is managing the customer experience. The following activities should be of prime importance—derive customer lifetime value, customer retention analysis, loyalty program management, and customer insights across channels.

Potential Use Cases

The key big data use cases for retailers are given in Table 13-4.

Table 13-4. *Big Data Use Cases for the Retail Industry*

Value Chain Phase	Potential Use Cases
Logistics and supply chain	Inventory and logistics optimization: Analysis of inventory data, replenishments in real time allows just-in-time inventory models to be pursued resulting in optimal savings for retailers. Tracking of shipments through RFID tags, GPS tracking of trucks and truck routes creates numerous opportunities for logistics optimization.
Merchandizing	Pricing optimization: Analysis of competitor prices, inventory, pricing, and point of sale (PoS) data in near real time creates opportunities for dynamic pricing based on market conditions and customer behavior.
Marketing	Promotion cost optimization: Analyze sales transactions for a period of promotion, historical trends as well as seasonal fluctuation to predict demand by geography and manage the promotion costs effectively.
Marketing	Promotion effectiveness through personalization: Integrating data such as customer buying propensity, market basket analysis, and geolocation data can help provide customized promotions to customers based on proximity to other related businesses.
Store operations	Labor optimization: Integration of labor data with PoS data and footfalls in a store can help create opportunities for labor optimization by store.
Store operations, supply chain, customer management	Fraud detection: By integrating store data, financial transaction data, product returns by customer and geography, browsing behavior, modern retailers can detect and prevent fraudulent activities that may occur across the value chain, such as PoS manipulation, employee theft, refunds, vendor issues, and stealing of financial data.

Emergence of Cloud-Based EIM Solutions

Cloud computing is transforming technology and offers a new way for enterprises to manage and utilize computing resources. Instead of making capital investments in purchasing, installing, and maintaining hardware and software, enterprises look to rent shared resources from a service provider and configure the service themselves. This model of computing significantly speeds up deployment times and lowers costs as well. There are multiple-deployment options including the following:

1. Public cloud—Application and computing resources are managed by third-party services provider. Public clouds offer enterprises financial savings as they don't need to invest in hardware and software, and they need to pay only for the usage or monthly subscription model. Public cloud adoption also helps free IT departments from focusing on software, hardware upgrades, and maintenance. There are high risks in this model including:

    a. Security and privacy—sensitive data resides outside the firewalls and can lead to huge financial losses in the case of data theft and hacking. More compliance needs—not being able to store sensitive customer data outside company firewalls, serve as bottlenecks to adoption.

    b. Costs—predicting costs and usage are difficult in a public cloud model.

    c. Outages—any outages in the public cloud can bring business critical applications to a halt.

2. Private cloud—Application and computing resources are managed by an internal IT team with some help from a third-party service provider. Although public clouds may be a faster option to deploy applications as well as provide cost savings, many enterprises adopt the private cloud model for catering to concerns about security and privacy of the data as well as application performance considerations. However, private clouds are a costlier option as they run on existing data centers and have their own costs to maintain.

3. Hybrid cloud—A hybrid architecture where certain critical data assets, components reside in an internal data center (private cloud) whereas other data assets, components are on a public hosted cloud. Many enterprises are adopting hybrid clouds to utilize the private cloud for bulk of processing needs and the public cloud to manage peak loads. There is a need for cloud management software that can cater to both public and private cloud environments. Some public cloud providers, such as Amazon's Elastic Cloud Compute (EC2), allow customers to have a virtual private cloud within the public cloud. This virtual private cloud can be connected to the enterprise's internal data centers using virtual private networks (VPNs). As is evident, hybrid clouds are complex to manage and maintain. According to a recent survey by RightScale in 2015, 82% of enterprises have a hybrid cloud strategy, up from 74% in 2014.

From a services standpoint, there are three classes of services for cloud computing. These are the following:

1. Software as a service (SaaS)—This delivers end user applications. These are leveraged to provide reports and dashboards to end users.

2. Platform as a service (PaaS)—This provides a development environment (programming languages, databases) allowing developers to build applications.

3. Infrastructure as a service (IaaS)—Where the service provider provides computing resources (servers, storage, and networking) that customers use to replace or augment existing resources.

Although enterprises initially looked at cloud deployment options for enterprise applications, now the focus is shifting to decision support systems such as information warehouses, data marts, and operational data stores. In the big data world, there also is discussion about putting data lakes on the cloud. I analyze some of the cloud deployment use cases in the information warehouses architecture.

Analytical Data Mart in the Public Cloud

The information warehouse is at full capacity and is unlikely to handle more workloads. Certain business units/departments have ad hoc query needs that are difficult to satisfy in the existing setup. In this scenario the analytical data marts can be placed on a public cloud to handle the ad hoc query loads. The cost of data transfers to the public cloud needs to be analyzed while making this decision (daily incremental feeds from information warehouse into data marts). Figure 13-5 shows the deployment use case.

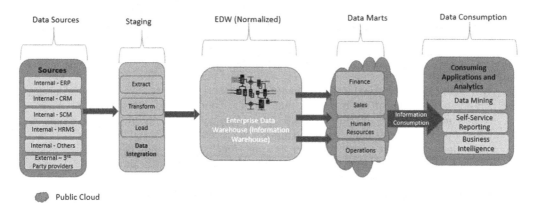

Figure 13-5. *Analytical data mart in the public cloud*

Analytical Data Mart and Business Intelligence Reports/Dashboards in the Public Cloud

In some instances certain departments/business units are looking for specific analytical data marts and reports/dashboards to be deployed and are skeptical about the speed of delivery from the IT function. In such instances it may be a good idea to have the analytical data mart and business intelligence (BI) reports/dashboards in the cloud. This is particularly useful in corporations in which there are multiple BI tools in use with no standards in place or in enterprises that are just embarking on the journey and are looking at faster turnaround times.

The benefits would be low cost deployment (due to the fact that they would not have to purchase BI software and only would pay based on a pay-per-use model) and a faster time to market. The challenges would be data transfer from the data warehouse (EDW) in the enterprise data center to the data mart in the public cloud. See Figure 13-6 for the deployment use case.

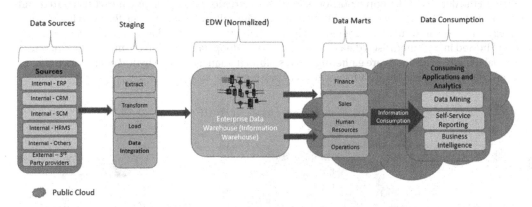

Figure 13-6. *Analytical data mart, business intelligence (BI) reports/dashboards in the public cloud*

Information Warehouse in the Public Cloud

If an enterprise, due to either its small size or technology driven, decides to have the information warehouse in the cloud, it would mean that the entire decision support landscape from staging, data integration, information warehouse, data marts, and business intelligence and analytics would be hosted on the cloud. The entire information warehouse would be a data as a service offering hosted with Amazon's Redshift, IBM's dashDB Enterprise MPP, or Microsoft's SQL Azure platform. Some challenges with this option would be data security and data transfers over the cloud. See Figure 13-7 for the deployment use case.

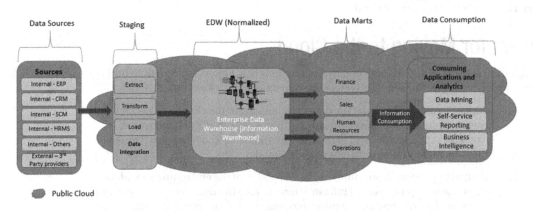

Figure 13-7. *Information warehouse in the public cloud*

Data Lake in a Hybrid Cloud

With the advent of big data solutions in enterprises, data lakes are coming into the data architecture. The data lakes often coexist with an information warehouse that stores the structured data while all unstructured data is stored in the data lake (Hadoop repository). In such a scenario, where there is a mix of structured and unstructured data as well as data created inside a company's firewall and external data, the need for a hybrid cloud strategy is felt. The information warehouse, which mainly stores internal data (structured), can be on a private cloud hosted in the enterprise's data center while the Hadoop data lake can be on a public cloud with all the external data. Hybrid clouds are gathering interest, though managing such hybrid environments is complex and challenging. See Figure 13-8 for a given deployment use case.

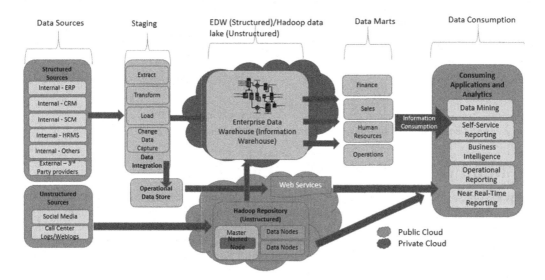

Figure 13-8. *Data lake in a hybrid cloud*

Drivers for Moving to the Cloud

Having looked at some of the common deployment options, it is important to analyze the key drivers to consider when moving the information warehouse or data lake to a cloud-based model.

1. Reduce IT costs—One of the key drivers for moving to the cloud is to cut capital investments made in buying software, hardware, as well as the IT personnel involvement in upgrades and maintenance of software and hardware. The cloud brings down these costs with a pay-per-use model.

2. Speed of implementation—Immediate availability of the environments with no delays due to procurement of infrastructure, reduce the timeline for deployment of information warehouse, and other information management solutions.

3. Demand elasticity—The leverage the massive computing resources available on the web, the scale up and scale down the compute resources based on the demand from business users, as well as ad hoc query needs.

4. Business focus on core areas—By outsourcing the information warehouse and BI applications, businesses can focus on core capabilities. This also enables IT resources more time for implementation and analysis work.

5. Data security—Data security, especially in industries, such as banking and health care, make it difficult to host data and applications on the public cloud. Even when enterprises go for a hybrid cloud care must be taken to ascertain which data resides on the public cloud and the data transfer mechanisms used for the internal data center to the public cloud.

6. Strong IT governance and service level agreements—These are needed when outsourcing information warehouses to cloud providers to ensure data availability, reliability of services provided, and so forth.

7. Vendor capabilities—Care must be taken while analyzing vendor capabilities as the market is full of multiple vendors with varying offerings and pricing models.

8. Data monetization opportunities—One of the impacts of moving to a cloud model is that the data becomes available to business partners that create monetization opportunities. For example, if a retailer moves the retail information warehouse to the cloud and shares the PoS data with consumer goods suppliers, it gives great insight about which products sell, in which geographies, and in which bundles.

Enterprises are increasingly looking at cloud deployment options and it's likely that this trend will continue given the nature of global businesses and greater need for data sharing and collaboration. Compliance standards also would need to be revisited to bring in the right controls for cloud-based models.

■ ■ ■

Glossary of Terms

The appendix covers the glossary of terms taken from individual chapters.

Chapter 1: Enterprise Information Management: Definition, Scope, and History

Business intelligence (BI) strategy—BI strategy deals with understanding the enterprise's business objectives and how the existing BI information landscape caters to the existing business information needs. BI strategy also derives the "to be state" based on future business objectives and helps to address the gaps in the information landscape. BI strategy helps to define the roadmap initiatives that would help an enterprise move from the current state to the "to be state."

Data architecture strategy—Data architecture strategy is one of the key pillars of the enterprise information management landscape. Data architecture defines the way data entities are modelled for system of record, data marts, and operational data stores. Data architecture includes the policies and rules concerning how data is sourced, stored, integrated, arranged, and used in decision support systems such as data warehouses, data marts, and operational data stores.

Data integration strategy—Data integration strategy deals with the optimal way in which enterprises can build the single consolidated view of business operations and performance. Data integration strategy looks at the best possible integration architecture and use of reusable integration components while integrating new sources of data. The objective of a data integration strategy is to ensure that the data integration architecture is optimal in terms of performance and scalability and can meet the enterprise batch window defined for processing data or real-time data needs.

Data security strategy—Data security strategy is crucial to protect the enterprise data assets that contain crucial information about enterprise business strategies, business performance, and intellectual property.

Data quality strategy—Often seen as an offshoot of information governance and master data strategies, data quality strategy deals with the processes and standards that ensure optimal quality for enterprise data assets over time; also will have strategies in place to monitor the quality of data over time.

Enterprise content management strategy—Enterprise content management is the strategies, methods, and tools used to capture, manage, store, and deliver documents and content related to organizational processes.

Enterprise information management (EIM)—EIM is a set of business processes, disciplines, and practices used to manage the information created through an organization's execution of business processes, managed by applications, and then this information is regarded by the enterprise as an asset.

Information governance strategy—Information governance strategy deals with classifying enterprise information assets based on usage patterns and business criticality and then defining governance structure and policies around the usage and consumption of the information assets. Information governance also brings into focus the quality of the information as the usage and decision making is often impacted by the quality of information rendered.

Master data management (MDM) strategy—Master data management strategy helps in delineating the source systems of master data creation and systems that would update/delete and consume master data. The MDM strategy deals with defining an MDM architecture that could be operational/analytical or hybrid depending on the business objectives the strategy needs to address. The MDM strategy also deals with building a business case for MDM that would highlight the tangible business benefits derived.

Chapter 2: The Lifecycle of Enterprise Information Management

Consumption—Consumption is the utilization of the information that is generated by a business process for 1) reporting/analytics or 2) as part of a business process or transaction execution need when the data is needed for a business decision.

Creation/receipt—Data is created at its point of origin in the business process. For instance, when a new customer is acquired a new customer identifier is assigned and relevant customer attributes are captured.

Data classification—Data classification is part of the information lifecycle management process that is used as a tool for classifying data that can be used by enterprises to answer questions such as: which types of data are available and whether the data is protected with the right controls and meets compliance needs as mandated by the industry.

Destruction/retire—When the archived data becomes outdated and not relevant to the current business processes and data consumption needs, the data can be classified as inactive. The data elements marked as inactive need to be destroyed.

Disposition/archival—At some point in the lifecycle of data there is no relevant use of this data element, the business decides it can now be marked for disposition or archival. In this step the data elements marked for archival are moved from the system of record or operational data stores to an offline storage mode such as optical disk or tape or even to a cloud-based storage model.

Distribution—Once the data is created, data is distributed to the relevant consuming applications that leverage the data to the running core business processes.

Information lifecycle management (ILM)—A set of policies, processes, practices, and tools used to align the business value of information with the most appropriate and cost effective IT infrastructure from the time of creation to the disposition phase.

Chapter 3: Components of Enterprise Information Management

Information architecture—Information architecture defines the blueprint for the information modeled to support business information and analytical needs.

Information delivery and consumption—Information delivery and consumption deal with a set of information delivery approaches and consumption styles through which the information processed from sourcing to various repositories can be consumed. The delivery of information is primarily through reports, scorecards/dashboards, activity monitoring, and so forth.

Information governance and quality—Information governance ensures that the enterprise data is trusted and its usage is governed based on the classification of the data being consumed and the rights of the requestor. Information governance is not about technology alone but about people taking responsibility for the information assets of their organization by looking at the processes they use to interact with information as well as how and why it is being used.

Information integration and exchange—Information integration and exchange is the process by which the information sourced is integrated into the information engine. In cases of information exchange with external systems or other internal systems that consume the data generated by the information integration process, there needs to be an interface agreement based on which the data is exchanged between the consuming system (subscriber) and publishing system.

Information models—Information models deal with how data is modeled to support decision support and analytical needs. There are multiple approaches for information modelling including top-down, bottom-up, and data vault.

Information sourcing—Information sourcing deals with the sourcing of data from a host of sources such as enterprise applications (e.g., ERP, SCM, MES, etc.) and other internal and external sources. The source data is extracted for a defined period and at a given latency defined by business analytical needs as well as availability of source data.

Information warehousing and reservoirs—These are key data repositories for supporting business performance monitoring and decision support including information warehouses, data marts, operational data stores, data lakes/data exploration zones, and so forth.

Master information management (MIM)—Master information management deals with the business processes, governance policies, standards, and technologies that consistently define and manage the critical master data entities, such as customer, product, and vendor of an organization, and provide a single version of truth. MDM ensures that on organization does not use multiple versions of the same master data in different parts of its business operations, which could result in higher costs of reconciliation.

Metadata management—Metadata is often defined as data about data and provides a context to the data it is associated with. Metadata can be managed through a set of defined processes wherein metadata is captured at each stage of a data management project. This ensures complete data lineage and traceability of data attributes as they move from information sourcing to information delivery and consumption.

Reference architecture—Reference architecture serves as a blueprint for all enterprise information management (EIM) solutions in an enterprise. Therefore, it can be seen as the big picture view of information management for a given enterprise. The reference architecture has all the relevant solution components needed to build an end-to-end EIM solution for the enterprise and includes the layers—information sourcing, master information management, information integration and exchange, information warehousing, and reservoir and information delivery and consumption.

Chapter 4: Pillar No. 1: Information Sourcing

Change data capture (CDC)—CDC is the process of capturing changes that were made in the source systems and applying these changes throughout the enterprise for both decision support systems such as information warehouse and operational data stores as well as other downstream consuming applications.

Full extraction—In full extraction, data from the relevant source system is extracted completely. This reflects the current and historical data available in the source system.

Incremental extraction—In incremental extraction, only the changes to the source data need to be tracked since the last successful extraction. Only the changes to the source data will be extracted (as delta changes) and loaded to the target consuming system.

Key performance indicators (KPI) dimension matrix—KPIs are used to measure key performance areas of a business. The KPI dimension matrix captures all relevant details concerning the KPI including functional area, granularity, and dimensions, whether it is based or derived, and so forth. KPI dimension matrix is the starting point for a source to target mapping exercise.

Operational data store (ODS)—ODS is the decision support database that integrated operational data from multiple source systems used to capture operational data and is used primarily for near real time operational reporting and analytics. ODSs are used to measure the operations processes efficiencies. The integration pattern is at the lowest levels of granularity and can happen from near real-time to multiple times in a day.

Pull mechanism—With the pull mechanism, the information integration engine is provided with the required access to the relevant source system tables and picks up the relevant data sets by querying the source system tables.

Push mechanism—With the push mechanism, source systems generate the source extracts that are moved through secure file transfer mechanisms to a file landing area. It is from the file landing area that these files are picked up for processing by the information integration engine.

Chapter 5: Pillar No. 2: Information Integration and Exchange

Data integration hubs—In a hub based model to data integration, data that is extracted from multiple sources flows through a centralized model (the hub) and is delivered from the hub to the consuming applications (spokes). In effect, the hub serves as a clearing house for data moving between the combinations of sources and targets. Data may flow through the hub on a scheduled, batch basis or in a real-time and granular fashion.

Extract information integration (EII)—EII is an integration mechanism by which heterogeneous data sources appear to the business users as single homogenous source. Some common approaches to this include data federation and data virtualization.

Extract, load, and transform (ELT)—In ELT, the source data is extracted and then loaded into the target database where the transformation to source data is accomplished using the database engine of the target database. The benefit of ELT is that the transformation happens inside the database and the transformed data does not have to be sent across the network as in the case with ETL approach.

Extract, transform, and load (ETL)—In case of ETL, the data is extracted from the source systems, transformed into a load ready form using the transformation/business rules, and then loaded into the target database.

Slowly changing dimensions (SCD)—Historical changes in dimension attributes, such as hierarchy changes, can be addressed through SCD. There are multiple types of SCD including Type 0, Type 1, Type 2, Type 3, Type 4, and Type 6 (hybrid SCD).

Real-time data integration—Real-time data integration deals with the integration of data from diverse source systems in near real time. The two approaches to near real-time integration are 1) change data capture (CDC) and 2) event or stream based integration.

Chapter 6: Pillar No. 3: Information Governance and Quality

Eurostat quality definition—The Eurostat quality assurance framework (QAF) is embedded in total quality management and describes the tools and procedures that have been put in place to ensure that the statistics produced are of high quality. The quality of statistical outputs is assessed against six criteria namely: relevance, accuracy, timeliness and punctuality, accessibility and clarity, comparability, and coherence.

Information governance council—The information governance council is the executive body who defines and approves the information policies and endorses the audit policies concerning information governance in an organization.

Information governance processes—Information governance processes help to classify information in terms of sensitivity, compliance, enterprise risk, and financial impact. They also help to develop the information consumption needs of business functions and units and provide a data access framework.

Information quality lifecycle—The end-to-end quality lifecycle for data assets in an enterprise include the following phases: information assessment, information cleansing, information enhancement, information consolidation, and continuous monitoring.

Information quality organization model—The information quality organization model is defined to ensure that the correct information quality processes are defined and followed in line with the governance framework in place. The organization model consists of strategic, tactical, and operational levels.

Information quality processes—Information quality processes are defined as part of setting up the quality framework that ensures that the quality of information sourced, integrated, transformed, and consumed within an enterprise is trustworthy, recent, consistent, and integrated.

ISO 8000, the international standard for data quality—The ISO 8000 is the ISO standard for data quality followed globally. It is comprised of the following parts: Part 1 (overview), Part 2 (vocabulary), part 100 (master data), and so forth.

Chapter 7: Pillar No. 4: Master Information Management

Architecture style—Architecture style is the style of implementation of the master information management. The four common styles are registry, consolidation, coexistence, and transaction.

Data integration—Data integration is one of the crucial design aspects of a master information management (MIM) solution as the integration of information into the hub happens through batch or real-time integration. Both batch and real-time integration can be handled through data integration tools and real-time integration also can be handled through enterprise application integration tools.

Data profiling—Data profiling is used to assess the existing state of data quality. It is also used to understand the duplicates in the master data or the gaps in linkages. It can be used to understand the scope of data enrichment to enhance the value of customer data assets.

Master information management (MIM) hub—MIM hubs serve as the single version of truth for all master data entities of the enterprise. MIM hubs can be built based on a custom model or an off-the-shelf tool (which provides a prebuilt model and can be further customized based on need). The hub needs to have audit capabilities to track changes made to the system as well integrate with existing enterprise information security requirements.

Chapter 8: Pillar No. 5: Information Warehousing

Data lake—Data lakes are repositories of raw source data in their native format that are stored for extended periods. Although Hadoop based repositories are gaining momentum to store raw data as their storage costs are low, often data discovery use cases also need structured reporting and analytics, which can be done out of an information warehouse thereby augmenting big data lakes.

Data vault—This is a hybrid modelling approach that uses part of Inmon and Kimball methodology. The data vault is based on the concept of hubs, links, and satellites. Hubs are master tables with source system keys (e.g., customer, product, location, etc.). Links represent associations/relationships between hubs with a validity period for these relationships. Satellites point to links and contain detailed attributes of associated transactions and their period of validity.

Information repository—An information repository is the actual store of enterprise data or the repository that houses the information warehouse. There is another repository in the information warehouse architecture namely the staging database. The staging layer typically retains data for a short period of time (can range from a few days to a month) depending on business requirements.

Information warehouse—The information warehouse is a system of record (SoR) that consolidates data from key enterprise applications as well as external data sources to give a 360 degree view of enterprise business functions. The SoR provides a data layer in which all enterprise data relationships are captured, providing a robust data platform for performing cross-functional reporting as well as building function specific analytical data marts (finance data mart, sales data mart, etc.).

Inmon or top down—This methodology involves modeling the information warehouse as an entity relationship model (3NF usually) and then designing the data marts as dimensional based on facts and dimensions.

Kimball or bottom up—This methodology involves modeling the data marts based on conformed dimensions and then building the data warehouse from the data marts using the dimension conformance.

Chapter 9: Pillar No. 6: Information Delivery and Consumption

Balanced scorecard—One of the widely adopted performance management frameworks is the balanced scorecard technique designed by Kaplan and Norton. Balanced scorecards involve looking at an enterprise (private, public, or nonprofit) through four perspectives: financial, customer, learning and growth, and operations.

Information delivery and consumption—Information delivery and consumption is a decision support system that consolidates discrete pieces of information from enterprise applications and other sources and provides management, operations managers, and administrative staff with a consolidated, consistent set of enterprise financial, operations, and performance metrics that help to enable decision making.

Performance management—Performance management measures the performance of a business unit or business through a set of key performance indicators. Performance management frameworks help to provide the key perspectives of performance and the associated metrics that need to be monitored and analyzed.

Planning and budgeting—With enterprises dealing in dynamic planning scenarios, the need for planning and budgeting applications that can handle dynamic planning scenarios due to unforeseen market conditions and also enable faster planning and decision-making cycles as compared to the traditional planning processes.

Scorecards and dashboards—Scorecards are used by enterprises to measure the progress against the enterprise strategy. Scorecards represent performance trends over a period of time such as monthly/quarterly/yearly; whereas dashboards indicate the status of a performance metric at a given point in time. In contrast, dashboards are used to represent actual granular data, they contain data that is more recent than that of scorecards.

Self-service business intelligence (BI)—A self-service BI is a semantic layer that enables business users to perform ad hoc reporting and analysis with no IT intervention. Self-service BI helps in the higher adoption of BI solutions.

Value driver trees—Value driver trees are defined as part of the performance management framework and help to link key result areas of a business to the actual key performance indicators, thereby providing a feedback mechanism into business performance.

Chapter 10: Pillar No. 7: Metadata Management

Business glossary—Business glossary like tools, serve as a dictionary for common business terms and measure definitions that promote the usage of data elements across business units and promote data collaboration in enterprises. The business glossary like tools enable business analysts, data analysts, and data stewards to work together to create, manage, and share the common understanding of business terms.

Metadata management—Metadata management is a discipline that deals with semantics and context of data as it is generated from a source system, through integration to information warehouses and data marts, delivery, and consumption, through channels to information consumers, and then to disposition and retirement. At each stage of the information supply chain, metadata or semantics and context about enterprise data is generated and captured through a metadata management solution.

Unified metadata approach—Unified metadata approach is the mechanism through which enterprises go about defining an unified metadata repository for all types of metadata that are part of the data generated from the enterprises business transactions and processes. Enterprises need to find a way to integrate these diverse metadata assets into a single repository to provide end-to-end data lineage capabilities, provide insight into the data as well as provide single and consistent definitions of key business data elements.

Unified metadata repository—The metadata solution must have a central repository to store all types of metadata—business, technical, and operational. The unified metadata repository must have a metamodel to store all the diverse types of metadata.

Chapter 11: Pillar No. 8: Big Data Components

Big data—Big data is a discipline that deals with processing, storing, and analyzing heterogeneous (structured/semistructured/unstructured) large data sets that cannot be handled by traditional information management technologies that have been used to process structured data. Gartner defined big data based on the three Vs: volume, velocity, and variety.

Data monetization—Analysis of big data in real time creates data monetization opportunities or new revenue streams for enterprises. For instance by analyzing weather data, retailers can provide customers with promotions in stores that are not impacted by weather events.

Data visualization—One of the key drivers for big data adoption is an ability to integrate and present both structured and unstructured data in the same report or dashboard. Big data visualization tools provide this ability as it enables end users to access data on both tablets, mobile devices, as well as through portals.

Internet of things (IOT)—IOT refers to a network of machines that have sensors and are interconnected enabling them to collect and exchange data. This interconnection enables devices to be controlled remotely resulting in process efficiencies and lower costs.

Shared data lake repository—A repository layer that is comprised of multiple repositories (information warehouse and Hadoop); it is often referred to as a shared data lake as it serves as a repository for all types of data—structured/unstructured, internal/external, traditional/new data sources.

Chapter 12: Building an Enterprise Information Management Solution

Enterprise information management (EIM) center of excellence (CoE)—EIM CoE is a shared services function that can address enterprise wide information management needs and provide fast, cost effective deployment of information management projects by linking people, process, and technology across the enterprise.

Enterprise information management (EIM) center of excellence (CoE) organization models—The common EIM CoE are classified as one of following models 1) decentralized, 2) hybrid, and 3) centralized.

Enterprise information management (EIM) blueprint—An EIM blueprint is a document that defines the basic information management principles of an enterprise and how these principles are derived from the business goals and objectives. The blueprint carries the business vision, which is mapped to business goals and objectives.

Enterprise information management (EIM) program governance framework—The EIM program governance framework ensures that there are governance mechanisms in place to check progress against the blueprint as well as provide overall governance to the EIM program.

Project prioritization—Project prioritization is the process in which projects are rated based on the project prioritization matrix defined by the core stakeholders, which in turn was based on criteria such as strategic alignment, business value, ease of implementation, availability of resource and requisite skills, availability of source system data, and so forth.

Chapter 13: EIM in Today's Business Environment

Exploration—Exploration of natural assets such as crude oil or minerals in natural habitats through exploratory drilling and seismic surveys.

Hybrid cloud—A hybrid architecture in which certain critical data assets, components reside in an internal data center (private cloud) while other data assets, components are on a public hosted cloud.

Private cloud—Application and computing resources are managed by an internal IT team with some help from a third-party service provider.

Processing—Production of crude oil and minerals from raw materials to finished/intermediate products. Mineral processing is the separation of commercially viable minerals from their ores.

Public cloud—The application and computing resources are managed by a third-party service provider.

Refining—Refining of crude petroleum to finished products such as naphtha, gasoline, diesel oil, LPG, and so forth.

Social master data management (MDM)—The integration of social media data concerning customer behavior, trends, and product discussions with structured master data entities to give a 360 degree view of master data entities is known as social MDM.

Streaming analytics—Stream processing enables enterprises to react to changing business conditions in near real time that enables applications such as fraud detection, trading, and system monitoring. Stream processing is when data streams or sensor data is processed (high event throughput versus number of queries) while complex event processing (CEP) utilizes event by event processing and aggregation with business rules and logic.

Index

Get the eBook for only $5!

Why limit yourself?

Now you can take the weightless companion with you wherever you go and access your content on your PC, phone, tablet, or reader.

Since you've purchased this print book, we're happy to offer you the eBook in all 3 formats for just $5.

Convenient and fully searchable, the PDF version enables you to easily find and copy code—or perform examples by quickly toggling between instructions and applications. The MOBI format is ideal for your Kindle, while the ePUB can be utilized on a variety of mobile devices.

To learn more, go to www.apress.com/companion or contact support@apress.com.